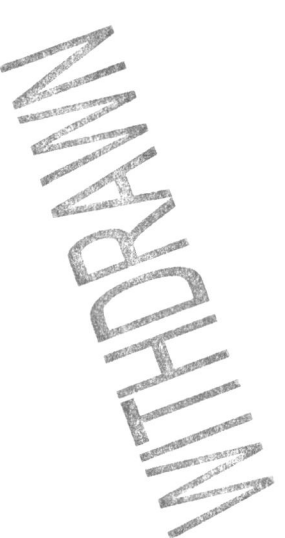

*The African Erosion Surface:
A Continental-Scale Synthesis of Geomorphology, Tectonics, and
Environmental Change over the Past 180 Million Years*

by

Kevin Burke
Department of Geosciences
University of Houston
Houston, Texas
USA

Yanni Gunnell
Department of Geography
University Denis Diderot Paris 7
CNRS UMR 8591
Paris
France

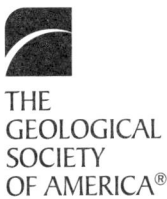

THE
GEOLOGICAL
SOCIETY
OF AMERICA®

Memoir 201

3300 Penrose Place, P.O. Box 9140 ▪ Boulder, Colorado 80301-9140, USA

2008

Copyright © 2008, The Geological Society of America (GSA). All rights reserved. GSA grants permission to individual scientists to make unlimited photocopies of one or more items from this volume for noncommercial purposes advancing science or education, including classroom use. For permission to make photocopies of any item in this volume for other noncommercial, nonprofit purposes, contact the Geological Society of America. Written permission is required from GSA for all other forms of capture or reproduction of any item in the volume including, but not limited to, all types of electronic or digital scanning or other digital or manual transformation of articles or any portion thereof, such as abstracts, into computer-readable and/or transmittable form for personal or corporate use, either noncommercial or commercial, for-profit or otherwise. Send permission requests to GSA Copyright Permissions, 3300 Penrose Place, P.O. Box 9140, Boulder, Colorado 80301-9140, USA.

Copyright is not claimed on any material prepared wholly by government employees within the scope of their employment

Published by The Geological Society of America, Inc.
3300 Penrose Place, P.O. Box 9140, Boulder, Colorado 80301-9140, USA
www.geosociety.org

Printed in U.S.A.
GSA Books Science Editors: Marion E. Bickford and Donald I. Siegel

Library of Congress Cataloging-in-Publication Data
Burke, Kevin, 1929–
 African erosion surface : a continental-scale synthesis of geomorphology, tectonics, and environmental change over the past 180 million years / by Kevin Burke, Yanni Gunnell.
   p. cm.
 Includes bibliographical references.
 ISBN-13: 978-0-8137-1201-7 (cloth)
 1. Geology, Structural—Africa. 2. Erosion—Africa. 3. Physical geography—Africa. 4. Morphotectonics—Africa. 5. Evolutionary paleoecology—Africa. I. Gunnell, Yanni, 1965– II. Title.
 QE635.B87 2008
 556—dc22

2008006912

Cover: The topographic basins and swells of Africa. Elevation range is from 0 to 5149 m above sea level, with lowest altitudes (basins) in purple to green through darker and lighter blue, and greater elevations (swells) in green to deep red through yellow and vermillion. Image shows Strahler stream orders 5 to 10 in growing order of line thickness. At this scale, no distinction can be made between perennial and intermittent streams, and due to the nature of the digital source data and behavior of the flow grid production routines, exactness is not guaranteed, particularly in internally drained regions such as deserts, rift valleys, and/or near large lakes. Drainage arteries are thus better defined here as topographic floodways than as streams, but in most places they highlight a clear patterning of modern drainage around active swells.

 Topography generated from 2-minute gridded Global Relief Data (ETOPO2 v. 2, U.S. Department of Commerce, National Oceanic and Atmospheric Administration, National Geophysical Data Center, 2006, http://www.ngdc.noaa.gov/mgg/fliers/06mgg01.html).

10 9 8 7 6 5 4 3 2 1

Contents

Abstract 1

The Unique Tectonic Identity of the African Continent: Expected Effects on the History
of Its Eroding Land Surface 2

General Characteristics of the African Surface 6

    What Is the African Surface? 6

    The African Surface: A Composite Surface 6

    Tectonic Evolution of Afro-Arabia and Its Controls on the History of the African Surface 7

    Structures That Influenced Local Relief on the African Surface during Its Evolution 10

    The African Surface in Basins That Experienced Marine Incursions between
    100 Ma and 34 Ma 11

Regional Characteristics of the African Surface 13

    The African Surface in Its Type Area: Southern Africa 13

        History of the Concept of the African Surface 13

        Recent Interpretations of the African Surface 15

        The African Surface in Southern Africa: A Summary 19

    West Africa: From Senegal to Cameroon 20

    Central Africa 24

    East Africa: The Great Swell and the Rift System 27

    Northeastern Africa and Arabia 29

    Summary: Regional Age Brackets for the African Surface 31

Bauxite Occurrences as Indices of the African Surface: Their Value and Limitations 31

    Bauxites … and Bauxites 31

    Examples of Orthobauxites and Laterites in Afro-Arabia Inherited from the Last
    Great Bauxite-Forming Interval, 70–40 Ma 32

    Staircases of Bauxite- and Laterite-Capped Surfaces in Their Key Area: West Africa 32

Bauxites in Africa That May Have Formed since the End of the Main Bauxite-Forming Interval, i.e., after 40 Ma 35

Synthesis on Bauxites and Laterites 36

The Imprint of Continental-Scale Paleoclimatic Variations in Afro-Arabia 37

Cretaceous Times, 140–65 Ma 37

Earlier Cenozoic Times, 65–34 Ma 38

Early Oligocene to Late Pliocene Times, 34–2.8 Ma 39

Late Pliocene and Quaternary Times, since 2.8 Ma 40

The African Surface in the Post–30 Ma Basin-and-Swell Reference Frame 40

Burial of the African Surface in Africa's Active Continental Basins 40

Topographic Uplift of the African Surface on Africa's Swells 40

Volcanism or Absence Thereof: Its Relevance to the African Surface 40

The African Surface on Africa's Swells: General Characteristics 42

The Great Swell of South Africa 44

Implications of the Cenozoic Swell Model for the History of the Great Escarpments at Africa's Continental Margins 46

Long-Lived Escarpments Close to Rifted Continental Margins? A Return to the Historical Type Area of the African Surface in Southern Africa 46

A New Perspective on the Great Escarpment of Southern Africa 52

Summary and Conclusions 57

Acknowledgments 59

References Cited 59

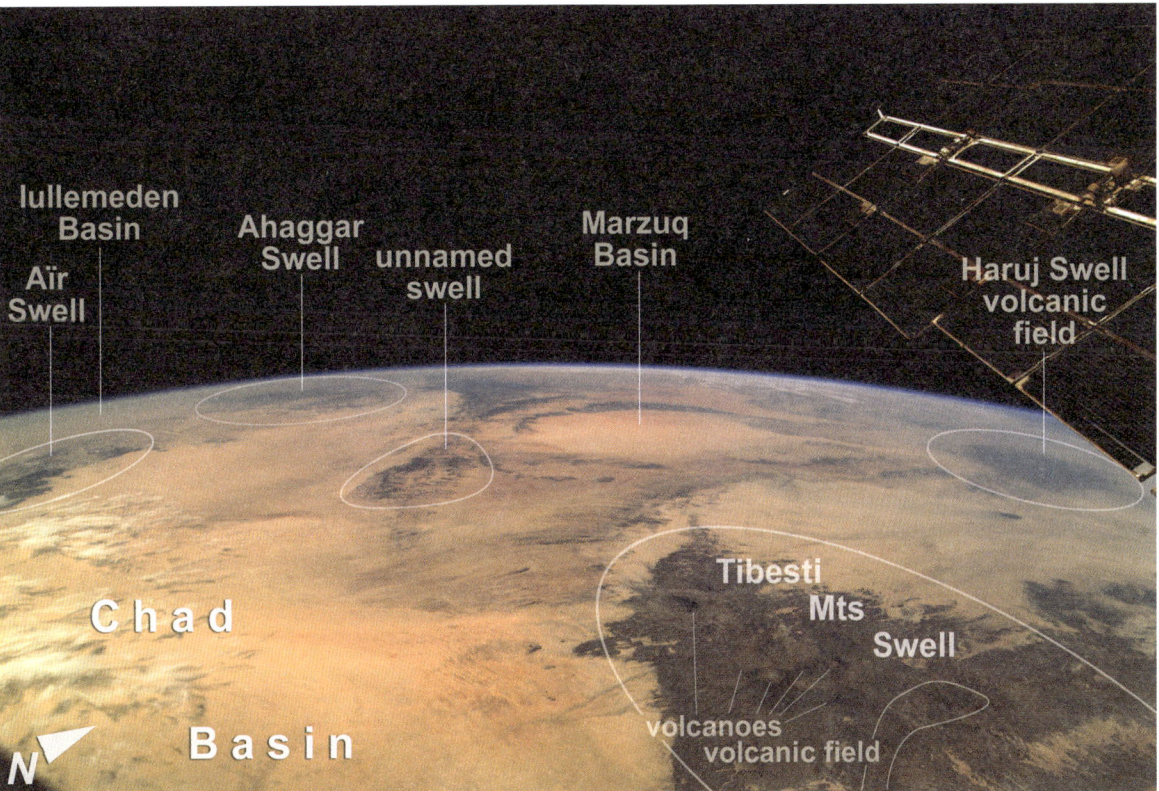

Africa's Rising Swells Volcano-capped swells and intervening basins of the central Sahara. Image taken from the International Space Station. Solar array panels top right. Provided and annotated by Justin Wilkinson. Image courtesy of the Image & Analysis Laboratory, NASA–Johnson Space Center (http://eol.jsc.nasa.gov). Unique image number ISS008E16999.

The Geological Society of America
Memoir 201
2008

The African Erosion Surface:
A Continental-Scale Synthesis of Geomorphology, Tectonics, and Environmental Change over the Past 180 Million Years

Kevin Burke*
Department of Geosciences, University of Houston, Houston, Texas, USA

Yanni Gunnell[†]
Department of Geography, University Denis Diderot Paris 7, CNRS UMR 8591, Paris, France

ABSTRACT

This outline of the topographic evolution of Africa tied to the history of the African Surface illustrates how a unique geomorphic history over the past 180 million years reflects the continent's distinctive tectonics. The African Surface is a composite surface of continental extent that developed as a result of erosion following two episodes of the initiation of ocean floor accretion around Afro-Arabia ca. 180 Ma and 125 Ma, respectively. The distinctive tectonic history of the African continent since 180 Ma has been dominated by (1) roughly concentric accretion of ocean floor following those two episodes; (2) slow movement of the continent during the past 200 m.y. over one of Earth's two major large low shear wave velocity provinces (LLSVPs) immediately above the core-mantle boundary; (3) the eruption during the past 200 m.y. of deep mantle plumes that have generated large igneous provinces (LIPs) from the core-mantle boundary only at the edge of the African LLSVP; and (4) two episodes during which basin-and-swell topography developed and abundant intracontinental rifts and much intra-plate volcanism occurred. Those episodes can be attributed to shallow convection resulting from plate pinning, i.e., arrested continental motion, induced by the successive eruption of the Karroo and Afar plumes.

Shallow convection during the second plate-pinning episode generated the basins and swells that dominate Africa's present relief. By the early Oligocene, Afro-Arabia was a low-elevation, low-relief land surface largely mantled by deeply weathered rock. When the Afar plume erupted ca. 31 Ma, this Oligocene land surface, defined here as the African Surface, started to be flexed upward on newly forming swells and to be buried in sedimentary basins both in the continental interior and at the continental margins. Today the African Surface has been stripped of its weathered cover and partly or completely eroded from some swells, but it also survives extensively in many areas where a lateritic or bauxitic cover has accordingly been preserved. Great

*Kevin.Burke@mail.uh.edu
[†]gunnell@paris7.jussieu.fr

Burke, K., and Gunnell, Y., 2008, The African Erosion Surface: A Continental-Scale Synthesis of Geomorphology, Tectonics, and Environmental Change over the Past 180 Million Years: Geological Society of America Memoir 201, 66 p., doi: 10.1130/2008.1201. For permission to copy, contact editing@geosociety.org. ©2008 The Geological Society of America. All rights reserved.

Escarpments, which are best developed in the southern part of the continent, have formed on some swell flanks since the swells began to rise during the past 30 m.y. They separate the high ground on the new swells from low lying areas, and because they face the ocean at some distance from the African coastline, they mimic rift flank escarpments at younger passive margins. The youthful Great Escarpments have developed in places where the original rift flank uplifts formed at the time of continental breakup. Their appearance is therefore deceptive.

The African Surface and its overlying bauxites and laterites embody a record of tectonic and environmental change, including episodes of partial flooding by the sea, during a 150-million-year long interval between 180 Ma and 30 Ma. Parts of African Surface history are well known for some areas and for some intervals. Analysis here attempts to integrate local histories and to work out how the surface of Afro-Arabia has evolved on the continental scale over the past ~180 m.y. We hope that because major landscape development theories have been spawned in Africa, a review that embodies modern tectonic ideas may prove useful in re-evaluation of theory both for Africa itself and for other continents. We recognize that in a continental-scale synthesis such as this, smoothing of local disparities is inevitable. Our expectation is that the ambitious model constructed on the basis of our review will serve as a lightning rod for elaborating alternative views and stimulating future research.

Keywords: tectonic deformation, continental denudation, sedimentary record, escarpment, landscape evolution, geochronology, rock weathering, climatic changes, paleodrainage.

THE UNIQUE TECTONIC IDENTITY OF THE AFRICAN CONTINENT: EXPECTED EFFECTS ON THE HISTORY OF ITS ERODING LAND SURFACE

The African continent measures 30×10^6 km², with an additional continental shelf area of 1.28×10^6 km² and a total coastline length of ~30,000 km. Africa's topography today reveals a modal elevation range of 0.4–0.6 km with a secondary peak around 0.8–1 km reflecting anomalous elevation, compared with other continents, of +0.2 to +0.5 km (Fig. 1). Addressing, as we do here, the possible causes and timing of the elevation changes that have generated this anomaly across an entire continent on one of the largest plates of the world is an ambitious task, lately undertaken for Africa at the continental and whole-plate scales by Summerfield (1985a, 1985b, 1996), Nyblade and Robinson (1994), Burke (1996), and Lithgow-Bertelloni and Silver (1998). Those studies have mostly addressed African topography by trying to understand how a deep-seated African mantle "superplume" might explain the peculiar topography (also called hypsometry) of continental Africa (Bond, 1979; Cogley, 1987). The large low shear wave velocity province (LLSVP) (Garnero et al., 2007) in the deep mantle below Africa is not a buoyant "superplume" but a structure in the deep mantle that has been stable for at least 300 m.y. Recognition of this fact has shown that explanations of topographic changes in Africa over the past 200 m.y. that relate those changes to changes in an underlying "superplume" are inappropriate (Torsvik et al., 2006).

The synthesis here is from the viewpoint that Africa's geologic, and especially its geomorphologic, evolution during the past 200 m.y. has been quite unlike that of any of Earth's other continents. This distinctive behavior has resulted primarily from four major tectonic controls: (1) new oceans were established around almost the entire continent in two episodes beginning ca. 180 Ma and 130 Ma (Fig. 1); (2) the continent has been nearly stationary for the past 200 m.y. over one of Earth's two major LLSVPs just above the core-mantle boundary (see Torsvik et al., 2006); (3) convective upwelling has occurred from the margins, but not from the interior, of that LLSVP and only from places where those margins are in contact with the core-mantle boundary, notably in plumes that were responsible for the eruption of Large Igneous Provinces (LIPs) (Burke et al., 2003a; Burke and Torsvik, 2004; Davaille et al., 2005; Torsvik et al., 2006); (4) African plate movement has been arrested with respect to the underlying sublithospheric mantle and Earth's spin axis since ca. 30 Ma (Burke and Wilson, 1972; McKenzie and Weiss, 1975; Burke et al., 2003a).

These four dominant tectonic controls have influenced the continental lithosphere and each other in a variety of ways. For example, the very small part played in African tectonic evolution by the slab-pull force, which is the largest of the forces that move lithospheric plates (Lithgow-Bertelloni and Richards, 1995) results from the opening of the Indian and Atlantic Oceans around the Afro-Arabian continent and a correspondingly limited role for subduction on Afro-Arabian plate boundaries. During the past 200 m.y., the African plate and its antecedent plates have largely escaped the pull of slabs of lithosphere in the mantle. As a result those plates have not moved very far across Earth's surface, so that Africa has rotated little with respect to the underlying convecting mantle and Earth's spin axis (Burke and Torsvik, 2004). A finite difference pole describing motion between 195 Ma and 30 Ma, since which time Africa has been in the position it now occupies (see, e.g., Burke and Wilson, 1972),

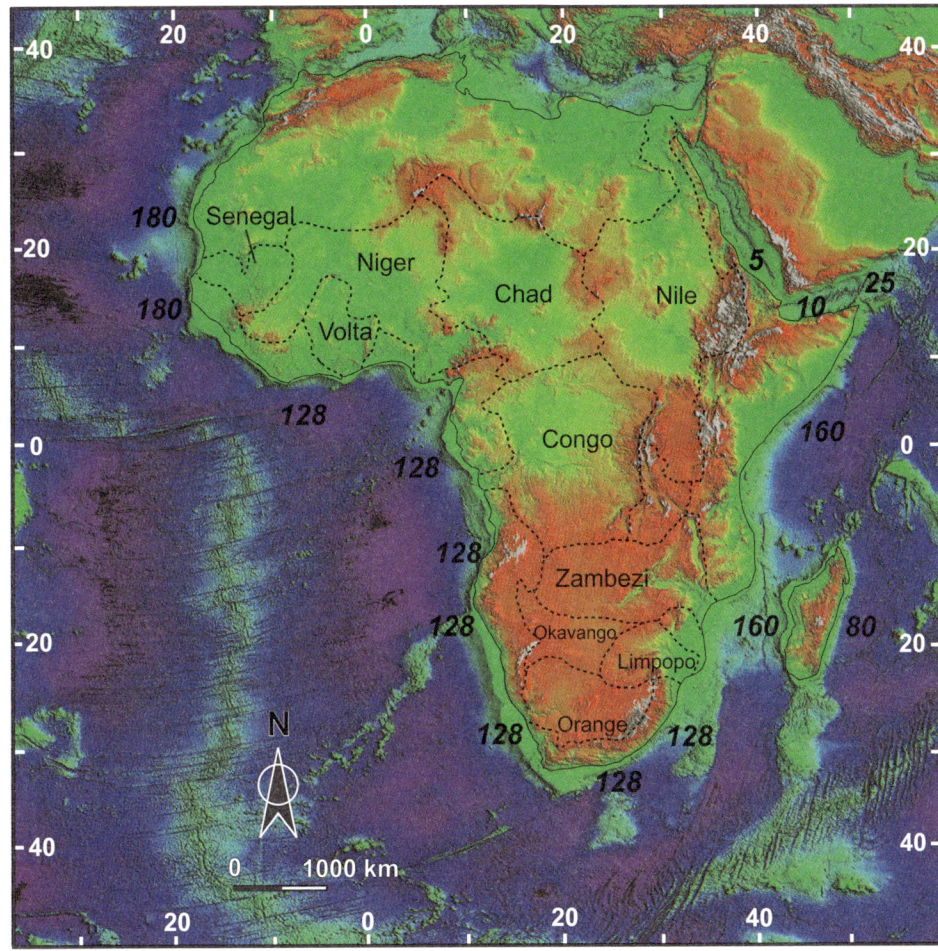

Figure 1. The African plate showing the major drainage basins of the continent. The coastline of Africa is shown as a thin black line. Offshore numbers indicate age (in Ma, requiring an error bar of ± 3 m.y.) of the oldest oceanic crust adjacent to the continental margins (from Burke et al., 2003a). Higher elevations shown in white and vermillion, lower elevations in yellow. Increasingly great depths below sea level are indicated by pale blue, dark blue, purple, and black. Green is used for both the lowest elevations on land and the shallower parts of submarine features, whether they are in great depths (for example, oceanic spreading centers and the Walvis Ridge), intermediate depths (such as the Mozambique Ridge), or shallow water (such as the continental shelves). Topography generated from 2-minute gridded Global Relief Data (ETOPO2 v. 2, U.S. Department of Commerce, National Oceanic and Atmospheric Administration, National Geophysical Data Center, 2006, http://www.ngdc.noaa.gov/mgg/fliers/06mgg01.html).

lies in the Atlantic Ocean. Counterclockwise rotation about that pole has moved the more eastern parts of the continent farther in latitude than the western parts. That has influenced the evolution of the land surface on the African continent by modifying the climate, and particularly the potential for extreme rock weathering and bauxite development. The availability of modern, highly resolved information on the timing of the eruption of mantle plumes and on the timing of the initiation of ocean floor formation, as well as the growing appreciation that the LLSVP under Africa is a compositionally distinct and stable body rather than a buoyant body capable of generating new elevation (see, e.g., van der Hilst and Karason, 1999; Torsvik et al., 2006), has launched us down the well-trodden but treacherous path worn by those who have tried to explain the nature and history of Africa's erosion surfaces. Establishment of the stability of the LLSVP has simplified our task because it shows that tectonic controls on changes in African topography reflect changes in near-surface tectonics and not deep-seated controls.

The driving tectonic hypothesis that underpins this study is that the history of the African continent since ca. 200 Ma has been punctuated twice by plume-induced plate-pinning episodes (PIPPEs) (see Burke et al., 2003a), resulting in arrested plate motion and enhanced vertical land surface instability. The rifts, basins, and swells (Fig. 2) set up in response to the shallow mantle convection (Fig. 3) developed during those intervals (Burke and Wilson, 1972; McKenzie and Weiss, 1975; England and Houseman 1984; Burke et al., 2003a; Li and Burke, 2006) and have governed the patterns, rates and styles of uplift and denudation across the continent. The first PIPPE was marked by the eruption of the Karroo plume that generated the Karroo LIP at 183 Ma (Eldholm and Coffin, 2000) (Fig. 4). Today, some of the rifts formed during the Karroo-PIPPE remain within Africa and Arabia (e.g., the West and Central African rift systems and the Yemen, Anza, Sirt, Abu Gharadiq, Shire, and Urema rift systems), whereas Atlantic-type or passive ocean margins have developed from other Karroo-PIPPE rift systems.

The Karroo-PIPPE ended at 133 Ma when the Tristan plume erupted and Residual Gondwana split into South America and Afro-Arabia. From that time the Afro-Arabian continent experienced a long (130–30 Ma) interval of relative tectonic quiescence. During these quieter times, the Afro-Arabian plate doubled in area as new ocean floor was formed at the spreading centers that almost completely surrounded the newly isolated continent. Between 130 Ma and 30 Ma, tectonic activity within

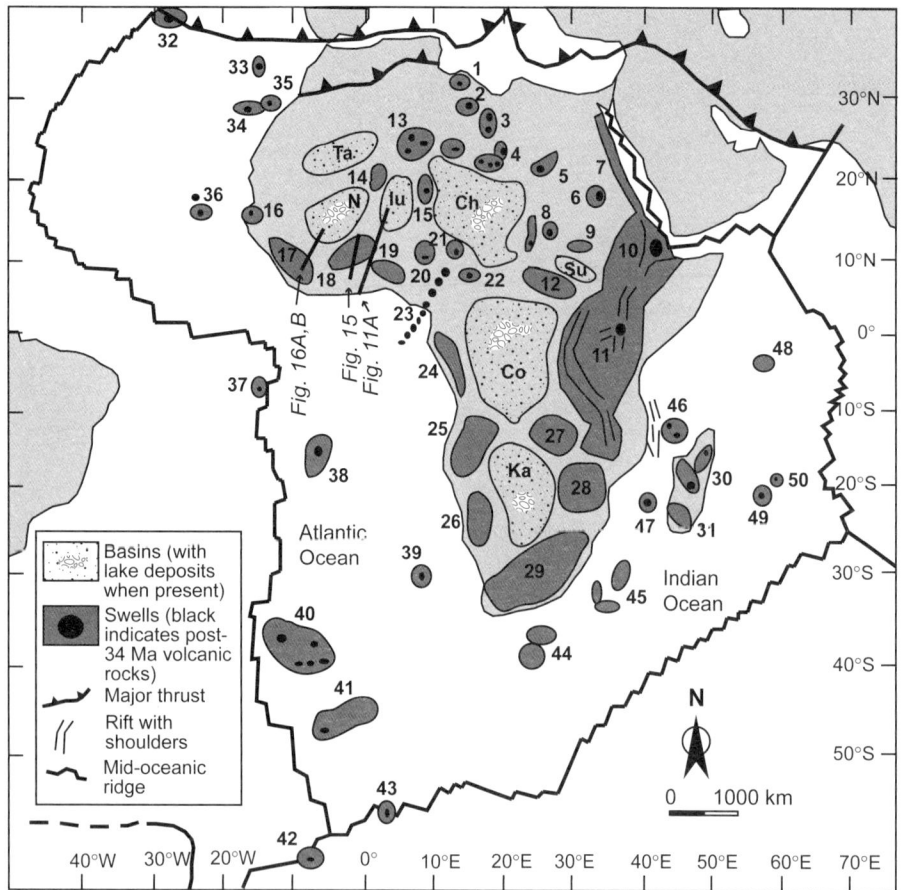

Figure 2. Basins and swells of Africa. All the basins and swells shown on this map began to form ca. 30 Ma, as the Afar plume began to arrest the African plate. This initiation date implies that only a small proportion of African relief today is older than 30 Ma. Offshore swells are difficult to identify and have been indicated only where they are capped by volcanoes, or where there is strong evidence of uplift (Seychelles). Cross-section lines in West Africa show locations of Figures 11, 15, and 16. *Swells:* 1—Tripoli; 2—Sawda; 3—Haruj; 4—Tibesti; 5—Uweinat; 6—Bayuda; 7—Red Sea Hills; 8—Darfur; 9—Alleira; 10—Ethiopian; 11—East African; 12—Nile-Congo; 13—Ahaggar (Hoggar); 14—Adrar; 15—Aïr; 16—Dakar; 17—Fouta Djallon; 18—Guinea (sometimes referred to as Leo); 19—South-West Nigeria; 20—Jos; 21—Biu; 22—Adamawa; 23—Cameroon Line (10 swells); 24—Mayombe; 25—Bie; 26—Namibia; 27—North Zambia; 28—Zimbabwe; 29—South Africa; 30—North and Central Malagasy; 31—South Malagasy; 32—Azores; 33—Madeira; 34—West Canary; 35—East Canary; 36—Cape Verde; 37—Ascension; 38—St Helena; 39—Vema; 40—Tristan; 41—Discovery; 42—Bouvet; 43—Shona; 44—Agulhas; 45—Mozambique Ridge; 46—Comores; 47—Atalante; 48—Seychelles; 49—Reunion; 50—Mauritius. *Basins*: Ta—Taoudeni; N—Niger; Iu—Iullemeden; Ch—Chad; Su—Sudd; Co—Congo; Ka—Kalahari.

the continent was episodic and was mainly concentrated in existing rifts. Those tectonic episodes, which locally involved intense folding and generated short-lived areas of high elevation, have been attributed to the propagation into the continent of stresses generated at the Tethyan continental margin by arc collisions (Burke and Dewey, 1974; Guiraud and Bosworth, 1997). The 100 m.y. interval of relative tectonic quiescence that lasted from 130 Ma to 30 Ma accommodated denudation, resulting in the formation of a composite, low lying surface of continental extent that is here called the African Surface.

The second PIPPE, which continues today, was initiated ca. 30 Ma, when the Afar plume erupted generating the Ethiopian LIP and pinning the overlying Afro-Arabian plate. That arrest led to the establishment of the active basin-and-swell structure of Africa (Holmes, 1944), to abundant intra-plate volcanic activity, and to the initiation of the presently active East African rift system including rifts on the future sites of the

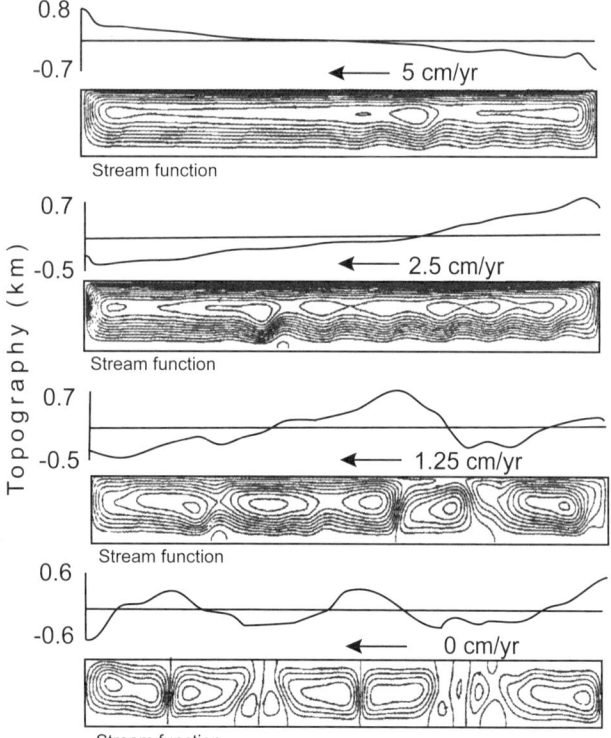

Figure 3. Shallow mantle convection under Africa (simplified from England and Houseman, 1984). On a moving plate, relief is low over the streamline flow. Basins and swells begin to appear on a stationary plate.

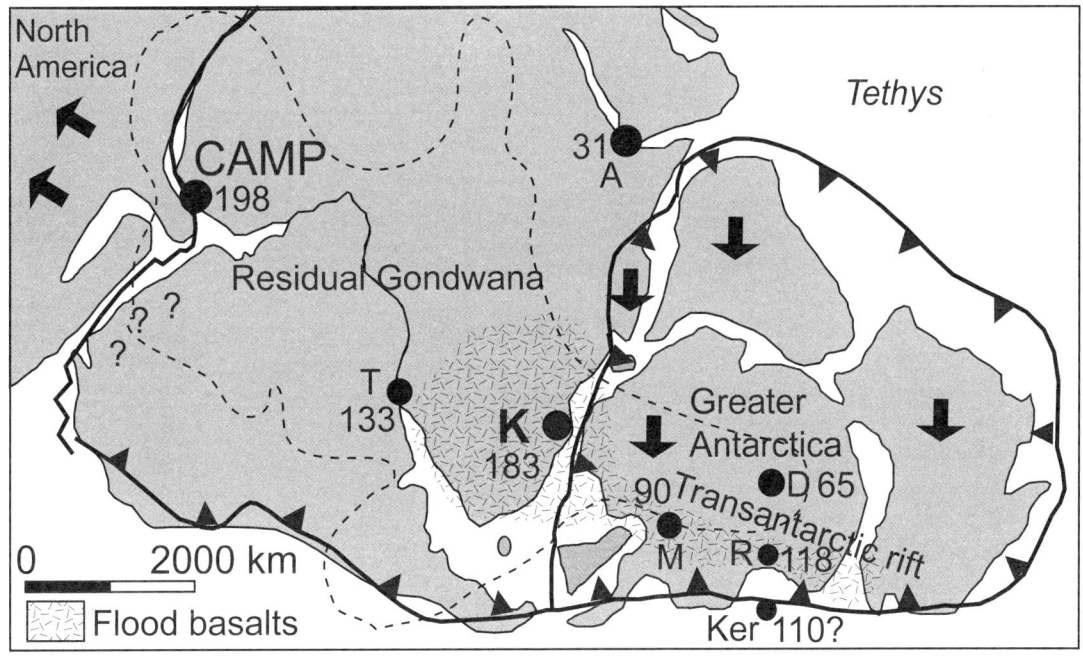

Figure 4. Sketch map of the African part of Pangea and neighboring areas as the supercontinent began to break up during a time interval straddling the eruption of the Karroo large igneous province (LIP) at 183 Ma. Pangea had been assembled ca. 300 Ma, and in areas remote from its cordilleran margin is likely to have become a generally low lying continent by 200 Ma. The African Surface began to exercise its dominance over residual Gondwana as soon as the rift shoulders associated with the departure of North America and greater Antarctica had been eroded away—possibly as early as ca. 150 Ma. Superimposed over Pangea is the dashed outline of a currently large low shear wave velocity province (LLSVP) near the base of the mantle (mapped tomographically [Grand et al. 1997]) and projected radially from its position under Africa. Eight plumes have erupted in the past 200 m.y. (ages given in Ma) and are considered to be of deep-seated origin because they all generated LIPs in just a few million years. The plumes are mapped with respect to Africa at their time of eruption: A—Afar; CAMP—Central Atlantic Magmatic Province; D—Deccan; K—Karroo; Ker—Kerguelen; M—Marion; R—Rajmahal; T—Tristan. Note that the eruption sites of these LIPS are all within ~1000 km of the edge of the projected LLSVP (Burke et al., 2003a). The edge of the LLSVP at the core/mantle boundary under Africa today was the source of all eight plumes, but only two (Karroo and Afar) succeeded in arresting the motion of the African plate effectively. The distribution of eruption ages of eight plumes between 200 Ma and 30 Ma indicates that the LLSVP has remained stationary relative to Africa and the spin-axis during the past 200 m.y. (Torsvik et al., 2006).

Red Sea and the Gulf of Aden (see review in Burke, 1996). The arrest of the African plate has been perturbed since ca. 5 Ma, when the Arabian, Somali, and Nubian plates were established. But rotations of those plates have been too small as yet to have modified structure and relief at the ~100 km length scale that we consider here.

In summary, two episodes of plate pinning, igneous activity, and associated basin-and-swell formation are thus distinguished from an intervening interval of tectonic quiescence. This simple threefold division of tectonic history provides a framework within which to analyze a highly complicated, diverse, and abundant literature on the erosion surfaces of Africa. Using tectonic analysis has diverted this study from tradition in a significant respect. Traditionally the denudational chronology of Africa has been reconstructed on the basis of altitude correlations (e.g., King, 1962, and many other studies cited here), or from the viewpoint of lateritic facies correlations tied to paleoclimatic reconstructions (e.g., Tardy and Roquin, 1998). Because the elevation of Africa's swells began within the past 30 m.y. (Burke, 1996; Burke et al., 2003a), the value of altitudinal correlations of erosion surfaces is limited to only the most recent past.

Although tectonics dominate the analysis in this paper, we include discussion of the influence of climatic changes on the evolution of the African Surface, emphasizing changes that followed the establishment of the Southern Ocean and the Eastern (or Great) Antarctic ice sheet ca. 34 Ma, as well as the changes that have followed the establishment of Northern Hemisphere glaciation and the closely linked initial formation of a desert in the Sahara at 2.8 Ma (DeMenocal, 1995).

The nature of the available literature requires this review be considered only a first modest step toward the probably unattainable holy grail of a thorough and integrated correlation of the myriad publications, reports, maps and other local studies that were produced during the twentieth century—mostly in English and French but also in other languages, by geological surveys, mining companies, engineers, academics, and a variety of research and development agencies. All of these studies are potentially useful in piecing together the complex and yet—as this contribution argues—remarkably coherent long-term geomorphic history of the African continent. Many of the studies carried out in various parts of Africa have remained piecemeal, unfinished, open ended, or hopelessly controversial because of diversity in methods and scope as well as for reasons related to the colonial balkanization of Africa and to political instability in some countries since independence. It is therefore true that the sum of a large number of local studies does not yet generate a coherent continent-wide picture. Some selectivity in interpretation is needed if a pattern is to be discerned. That degree of selectivity has been exercised here.

GENERAL CHARACTERISTICS OF THE AFRICAN SURFACE

What Is the African Surface?

The African Surface, however variously and vaguely defined, has long been considered the most extensively preserved ancient erosion surface of the African continent (see, e.g., King, 1962). That dominance has become even clearer with the emerging consensus (e.g., Partridge and Maud, 1987) that pre-Jurassic (pre–ca. 200 Ma) erosion surfaces are rarely if ever preserved in Afro-Arabia except in places where they have been exhumed from beneath a sedimentary rock cover. The African Surface is not only the most extensive but also the oldest major erosion surface of the continent.

The following sections embody reviews of what is known about the African Surface, referring to many classic studies and suggesting answers to such questions as What is the African Surface? What is the present distribution of the African Surface both geographically and in elevation? When did the African Surface form? Is it a monogenetic or a polygenetic surface? To what extent can its paleoelevations be reconstructed? Was the development of the African Surface related to a particular set of climatic circumstances? Did those circumstances operate at a particular time or times? Does the African Surface show distinctive weathering features? Do these lie in the extent, the thickness, or the composition of weathered rock? What does African Surface distribution tell us about relatively recent African tectonics? Is the African Surface recognizable when buried beneath younger sedimentary rocks in the subsurface, and what makes it recognizable? Is the African Surface intercalated within marine sedimentary successions? In places where the African Surface is preserved at Earth's surface is it still experiencing active weathering? Are there marked regional differences in the character of the African Surface, e.g., weathering mantles or soil covers, among the various regions of Africa?

The African Surface: A Composite Surface

In regions of continental extent, erosion surfaces have resulted from the destruction of rugged topography including mountain belts, rift-flank uplifts, and volcanoes. The time scales involved imply that these land surfaces are likely to have been affected by numerous climatic cycles and by a wide, often unspecifiable range of surface processes. The task of determining the age of an erosion surface is therefore one of the longer-standing conundrums of geomorphological research. Definitions can, perhaps, help.

In the light of practices that culminated during the 1950s, when disciples of W.M. Davis were apt to see flat surfaces everywhere and multiply erosion cycles on the basis of statistical inferences from maps (e.g., Marker and Brook, who in 1970 identified eight different "surfaces" in South Africa based on a statistical elevation analysis of five 1:50 000 topographic sheets), we recognize that caution is appropriate in interpreting the African landscape. Although studies such as those of Marker and Brook may have been partly responsible for ushering in the era of the rejection *en bloc* of Davisian geomorphology spearheaded by Chorley (1965), the position of some rejectionists who contend that erosion cycles and their resulting planar surfaces are articles of faith based on untestable hypotheses appears too extreme: extrapolating topography into the sky (and, in doing so, substituting space for time, into the past) is not a priori exposed to greater error than inferring the geometry of geologic structures at depth in the crust, lithosphere, and mantle—a pursuit to which geologists routinely commit their careers. The approach in this paper is considered to be intermediate between extremes. It is one of "lumpers" rather than "splitters" in the reconstruction of the stages of long-term landscape development across the African continent.

A great deal of confusion has surrounded the local mapping and age-bracketing of the African Surface. Dated cover rocks are present in some areas, as in Kenya (see Saggerson and Baker, 1965), and not in others, as for instance in neighboring Uganda. Controversy around the age of erosion surfaces has often hinged on the multiplicity of ages established by local studies and the failure to look at the bigger picture. One reason for this is that many, if not all, major planation surfaces are *composite*—one of the major underlying principles of this study. Dumont (1991) pointed out that any given erosion surface carries several ages, each with its own definition: (1) A surface can be defined by its *initial age*, i.e., the time at which a key lowering in base level triggered the onset of a new erosion cycle and the reshaping of pre-existing land surfaces. (2) An erosion surface can also be defined by a variety of *local ages*. This is possibly the most common approach in studies of denudation chronology, in which surfaces are time-bracketed by the age of the oldest datable deposit (sedimentary or volcanic) that seals them. At the continental scale, this age is likely to vary considerably, as indeed it may at the regional level. This has not deterred geomorphologists such as L.C. King from constructing global schemes of denudation chronology involving the extrapolation of locally dated surfaces over great distances. (3) A third kind of age can come from the expansion of drainage basins related to a change in base level that terminates the stability of the eroded land surface. The time of onset of this rejuvenation process defines the *terminal age* of a surface. This terminal age is effectively also the initial age of a younger land surface in the making.

A surface is polygenetic or composite, as the term is used in this paper, (1) if it evolved over a long period of time, typically 10^7 to 10^8 years, and (2) if during that time it experienced—following the eustatic analog of Haq et al. (1987) in a different although related field of study—second-order environmental variations (typically climatically or tectonically controlled) that induced changes in the way that the surface developed. For example, because of a short-term climatic change from wetter to drier, a pediment gravel might be deposited on a surface that had been experiencing intense chemical weathering. With a reversion to the earlier climatic conditions, chemical (for instance lateritic) weathering could resume. In a detailed local study, three distinct surfaces might be distinguished, but in a more regional study the first and second laterites and the intervening pediment gravel could all be considered to represent locally controlled developments on a composite surface. In a second example, a surface near sea level that was experiencing deep lateritic weathering might be flooded in a marine transgression and experience the deposition of a thin sedimentary sequence, perhaps consisting dominantly of carbonate sedimentary rocks. With the retreat of the sea, deep lateritic weathering could resume. Distinct surfaces might be discriminated in a local study but the whole could be considered a single surface in a regional study. In a third example, a composite surface might be recognized as a single surface above and below a topographic step controlled by a lithological contrast. A large dolerite sill or a sandstone escarpment might cause the step in the topography. Surfaces capped by lateritic crust of the same kind and formed at the same time above and below the step are considered in this review to be elements of the same composite surface. In Africa, the height of such steps is typically 100 m or less and seldom greater than 200 m.

In the light of this approach, and given that a maximum age of ca. 180 Ma was the time of Pangean breakup events on the bulge of Africa and on the Indian Ocean coast of the continent, "African Surface" is the name given here to the composite erosion surface that dominated Afro-Arabian scenery ca. 30 Ma as the outcome of up to 150 m.y. of continental denudation. A second breakup episode ca. 125 Ma, when the South Atlantic began to open, provided the second of two initial ages for the African Surface; and 30 Ma, when Africa's swells began to rise, defines the surface's terminal age. The African Surface inevitably has a range of local ages in various parts of Africa depending on which bracketing chronostratigraphic clues are locally available.

As in this study, Partridge and Maud (1987) in their review of South African erosion surfaces emphasized the composite character of the African Surface. They attributed the development of the African Surface, in its South African type area, to complex cycles of erosion during an interval similar in duration to that considered here. For Partridge and Maud, the interval extended from a "Gondwana breakup rifting event," which they placed ca. 150 Ma, to the early Miocene (ca. 20 Ma). The difference between their estimated duration for African Surface development and ours is small and certainly not, given the present limits of temporal resolution, critically important. It probably arises because Partridge and Maud (1987) confined their consideration to southern Africa whereas we consider the whole of Afro-Arabia.

Thin deposits of Cretaceous nonmarine sedimentary rocks, for example those preserved in the Congo Basin as well as those making up the "Nubian" sandstone (quotation marks indicate that use of the term for a formally defined stratigraphic unit has long been abandoned) and the other rocks of the *continental intercalaire* and the Hamadien of the Sahara and Sahel, illustrate the composite nature of the African Surface. They lie above one of several locally mappable surfaces that together make up the composite African Surface and are themselves overlain by another such surface. Those thin Cretaceous accumulations can be thought of as deposits lying on the African Surface; they can also be viewed as thin sedimentary lag units correlative of one phase of relief reduction in a part of Africa and therefore an integral part of the weathering and stripping cycles that produced the African Surface. Except near the continental margins and in places where they have been deposited over ancient rifts, the rocks deposited on the African Surface are thin, typically <100 m thick. The nonmarine sedimentary rocks that represent episodes of deposition during the mainly erosional development of the African Surface in the interior of Afro-Arabia are not only thin but also only locally preserved. For that reason it is usually only close to the continental margins and to rift shoulders, which sustained greater local crustal uplift, that individual surfaces recognizable as separate elements of a composite African Surface can be distinguished.

Tectonic Evolution of Afro-Arabia and Its Controls on the History of the African Surface

Beginning from the time during the Toarcian ca. 180 Ma, when the Central Atlantic and Indian Oceans began to open, erosion reduced Afro-Arabia to a low elevation. During Late Jurassic and Early Cretaceous times (150–100 Ma) the shoulders of the newly formed rifted continental margins around Afro-Arabia were eroded to low elevations. By ca. 100 Ma in late Albian times, the whole continent had become low lying, probably not exceeding mean elevations of 0.5 km a.s.l. (above sea level) according to Sahagian (1988) and Guiraud and Bosworth (1997). Between 100 Ma and 30 Ma, and in some regions starting as early as 150 Ma, the low lying Afro-Arabian continent was shaped by the slowly evolving composite erosion surface to which the name "African Surface" is here applied. In some places individual surfaces that contribute to making up the composite African surface can be distinguished, but in many areas such separate surfaces cannot be, or have not yet been, mapped. Although the recognition of individual surfaces that contribute to the composite African Surface is necessary in local and regional studies, it is essential to recognize the subordinate character, on the continental scale, of all those surfaces to the great composite

African Surface. From ca. 100 Ma, rivers flowing from the interior toward the continental margin breached the eroding shoulder uplifts, and wedges of siliciclastic sediment propagated onto the carbonate platform at the continental margin. Examples on the South Atlantic margins of Africa include the Cretaceous Orange River (Jungslager, 1999, especially Figs. 2 and 14 therein), the Cretaceous Congo River (Karner and Driscoll, 1999, p. 290), the Cretaceous rivers of the Kwanza Basin (Brognon and Verrier, 1966), and the Cretaceous rivers of Gabon (Brink, 1974; Teisserenc and Villemin, 1989). The timing of the change from carbonate to siliciclastic rocks for basins at the margins of Afro-Arabia is set down in Figure 5. The timing is best defined in tropical latitudes, where carbonate rocks are well developed. In higher latitudes, where there may have been little or no carbonate rock deposition, the elimination of the rift shoulders and the rapidity of change to siliciclastic deposition at the time when rivers began to cut through the rift shoulders is not so clear. The first development of a delta prograding onto the continental shelf may be the best indicator of the completion of rift-shoulder degradation in those regions. Evidence from the conjugate shore of the ocean can also be informative (see, e.g., Coward et al., 1999).

The westward bulge of Africa and Afro-Arabia's east coast as far south as Durban had become low lying by ca. 150 Ma (e.g., Coffin and Rabinowitz, 1988, for the east coast; Jansa and Weidmann, 1982, for the central Atlantic). Africa's Gulf of Guinea, South Atlantic, and Mediterranean coasts, apart from that in the Maghreb where there was almost continuous tectonic activity, had become similarly low lying by 100 Ma. Some of the intracontinental rifts that had been initiated in the interior of the Residual Gondwana continent between 180 and 130 Ma (Fig. 6) (Burke et al., 2003a) came to host the major rivers of the continent. Those rift valleys continued to receive sedimentary deposits during later intervals of Afro-Arabian continental evolution. For example, those of the Sirt Basin in Libya experienced the deposition of several kilometers of sedimentary rocks long after they had been initiated, but no new rifts were initiated on the Afro-Arabian continent between 130 Ma and 30 Ma (Burke and Whiteman, 1972; Burke, 1996; Burke et al., 2003a).

Intracontinental rifts initiated between 180 Ma and 130 Ma experienced episodes of later faulting, folding, and basin inversion in a succession of events propagated into the continent from the Arabian coast and from the Maghreb by the Tethyan convergence (Badalini et al., 2002). The most prominent was an intense and short-lived continent-wide Santonian deformation episode at 84 Ma (Fig. 7) that has been attributed to the collision of Arabia with a Tethyan arc (Burke and Dewey, 1974; Guiraud and Bosworth, 1997; Burke et al., 2003a). In summary, most of the area of Afro-Arabia situated away from intracontinental rifts and rifted margins had become tectonically quiet by the end of Albian times (ca. 100 Ma); i.e., no new rifts were formed between ca. 130 and 30 Ma. This did not prevent persistent rift activity (i.e., subsidence, deposition, and rock deformation) from continuing in some of the rifts that had begun to form before 133 Ma, particularly in regions that were situated closest to the Tethyan collision

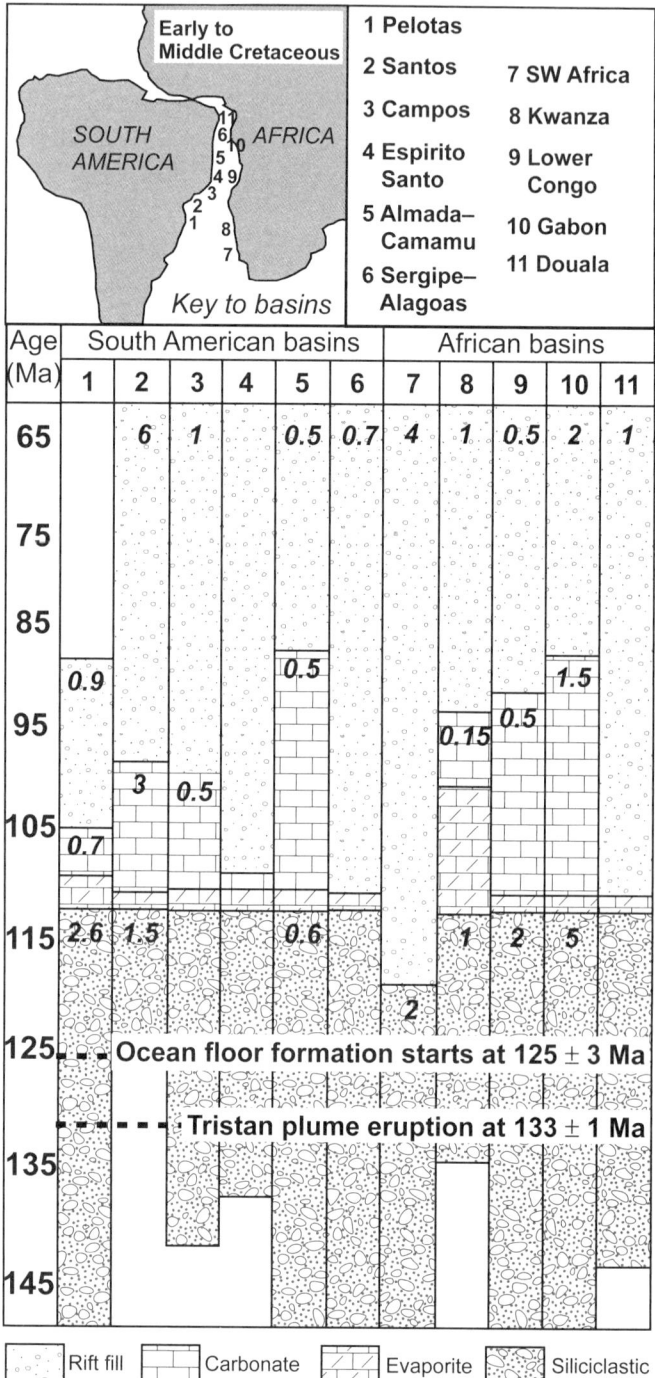

Figure 5. Limited longevity of primary rift shoulders at South Atlantic rifted margins inferred from offshore sedimentary sequences. Rift deposition persisted in rifts formed at 145 ± 5 Ma and on ocean floor under air until seawater spilled in to form evaporites. After evaporites were deposited, carbonate deposition followed, ending 20–35 m.y. after ocean floor began to form. Rift shoulders had been eroded by the time siliciclastic deposition began at 105–90 Ma. Rift-shoulder escarpments existed for less than 50 m.y. Deposit thickness (numbers in italics) is in kilometers. Based on stratigraphic data in figures of Coward et al. (1999).

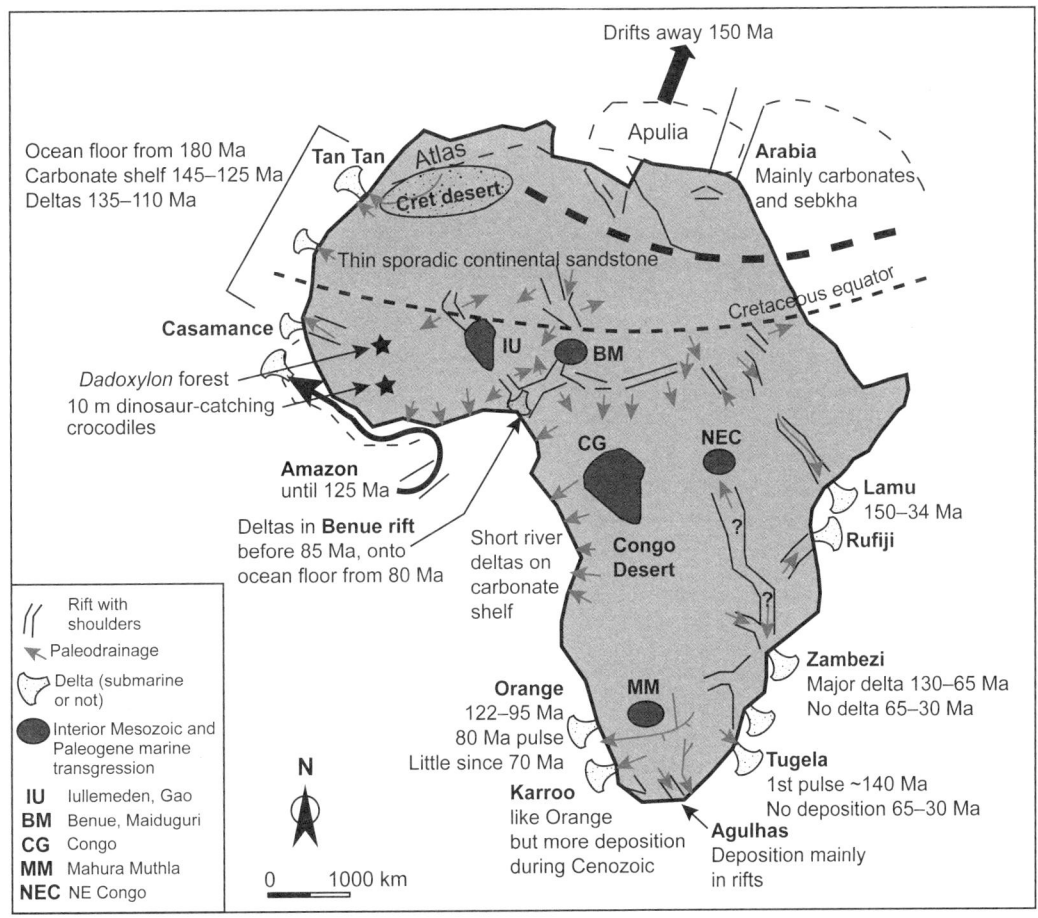

Figure 6. Schematic map of African rifts, rivers, and deltas ca. 180–30 Ma (Santonian tectonics at 84 Ma excluded). Laterites and bauxites formed between desert latitudes, silcretes and calcretes farther south. Cretaceous and Paleogene (100–40 Ma) marine sedimentary rocks preserved in the interior of Africa indicate the generally low-lying character of the continent. Five Mesozoic and Paleogene marine transgressions occurred in the Iullemeden, Benue, and Maiduguri Basins between 100 and 50 Ma. The NE Congo Basin recorded only Late Cretaceous transgressions. Non-marine land plants in river gravels have been found in the Mahura Muthla region. The Congo Basin contains shallow-water Cretaceous marine strata supposed to have connected with the trans-Saharan seaway in the north rather than with the expanding Atlantic Basin in the west (Giresse, 2005). North of heavy-dash line the northeast Afro-Arabian continent was subjected to rapid shoreline fluctuations between 150 and 30 Ma. The Afro-Arabian continent formed ca. 125 Ma (end of Barremian times) when South America departed from residual Gondwana. Apulia left from the north coast of Africa a little earlier. Once the new rift shoulders were eliminated by ca. 100 Ma, drainage was concentrated in a small number of rivers mainly flowing in rifts formed between 180 and 140 Ma, most of which represent failed arms of triple rift systems. Some rivers (e.g., the Zambezi) flowed in reactivated older rifts. From 100 Ma until 30 Ma the continent was generally low lying, its mean topography becoming progressively smoother. It was dominated by the African Surface and drained largely by a small number of large rivers. The deltas of most of these are shown on this map. Of the major rivers, only the Tan Tan, Orange (also called Kalahari), and Karroo Rivers did not flow in rifts. The heavy-dash line across northern Africa illustrates the position of the Cretaceous and pre–30 Ma Cenozoic shoreline, which lay far into the continent at highstands and retreated to the present coast during lowstands. Rifts of the Sirt Basin contain a record of those transgressions and regressions.

zone—for example, the Sirt Basin rifts of Libya, the Muglad rift of Sudan, and the Benue rift of Nigeria, which accommodated several kilometers of Cretaceous to Paleogene sedimentary rocks and underwent folding and faulting during occasional events such as the Santonian episode.

The relatively quiet tectonic setting of Afro-Arabia, which dominated the great continent from 130 to 30 Ma, provided an environment for the development, over tens of millions of years, of an extensive, deeply weathered erosion surface at low elevation. In Afro-Arabia north of the equator, as explored in further detail in the regional overviews provided below, there is abundant evidence of flooding during highstands of the sea. Farther south, at least in part because of erosion during the greater elevation of the post–30 Ma swells, evidence of marine flooding is scarcer.

This represents a critically distinctive characteristic of the African continent because comparable tectonic conditions over this long an interval have not been common on the larger continents of the earth in the geologic past. Most continents have carried a cordilleran mountain system on one coast, as North and South America do today. Two comparatively small continents

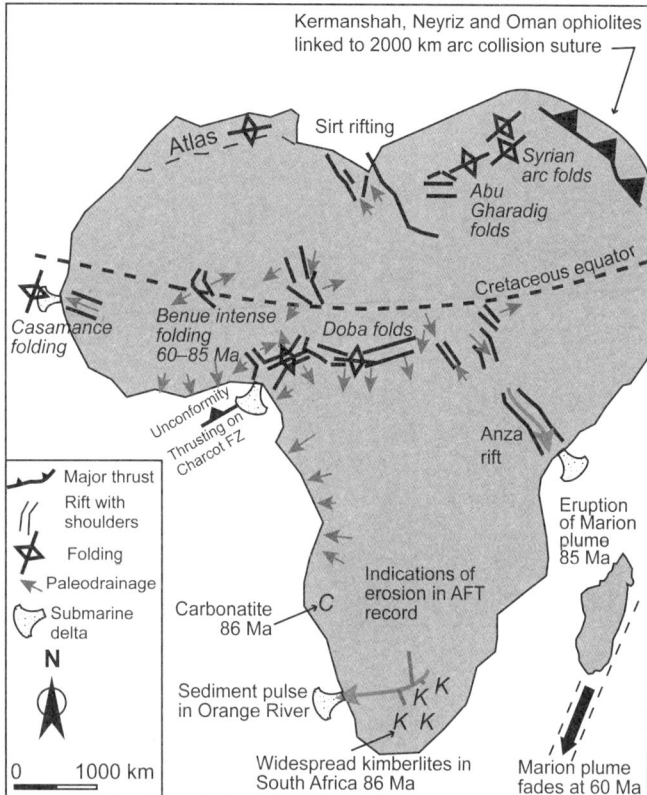

Figure 7. Schematic outline of some tectonic, erosional, depositional, and igneous events related to the Africa-wide tectonic pulse in Santonian times (83 ± 2 Ma). Products of this Santonian continent-wide relief-forming event were rapidly eroded (Guiraud and Bosworth, 1997). Gray arrows indicate drainage directions and loci of continental deposition as high ground was eroded away during a short pulse lasting probably not more than 10 m.y. (see, e.g., Fig. 2 in Jungslager, 1999).

provide relatively recent analogs of the isolated Afro-Arabian continent. One is Australia for the interval between ca. 80 Ma, when the Tasman Sea began to form, and 20 Ma, when collision with Indonesia began. The other is India between 135 Ma, when it separated from Antarctica, and 50 Ma, when the Indo-Asian collision began. On both these continents, as across Africa, extensive surfaces of low relief developed. On the Indian continent, those surfaces became modified first when the Deccan LIP was erupted at 65 Ma, and later by deformation after the Himalayan collision at 50 Ma, particularly in the past 10 m.y. In the Australian continent, deformation was concentrated in New Guinea after the collision with Indonesian arc systems. On the western side of Australia, deformation of Earth's surface has been minimal but elevation of <2 km close to Australia's east coast within the past 30 m.y. has been related to widespread and substantial basaltic eruption. In the more remote past the much larger (25 × 10^6 km^2) continent of Laurentia appears to have been completely surrounded by Atlantic-type rifted margins between 600 Ma and 450 Ma. It may provide an analog of Afro-Arabia between 125 and 30 Ma, although Laurentia was largely flooded by the sea during the interval in which it was surrounded by Atlantic-type continental margins.

Structures That Influenced Local Relief on the African Surface during Its Evolution

By analogy with the western two-thirds of Australia today, there is unlikely to have been much relief on Afro-Arabia greater than ~0.5 km above common pediment levels. Most elevations on the African Surface probably rose no more than 0.2 km above a pediment surface. Existing areas in which the African Surface is exposed show that among factors that have influenced local relief, the exhumation of gently inclined surfaces has been important. For example, the African Surface in the Gauteng and North West Provinces of South Africa (near 25°S, 28°E, formerly the central Transvaal) is remarkably flat where it lies on the exhumed outcrop of the Ecca (Permian) shale. Other exhumed flat surfaces include the sub-Cambrian unconformity in the Sahara (Beuf et al., 1971; Klein, 1997) and the sub-Dwyka surface of southern Africa, which is controlled by the contrast in resistance to erosion between underlying, mostly crystalline, rocks and overlying sedimentary rocks. The sub-Dwyka surface is quite flat except where, locally, glaciation has cut deeply (de Wit, 1999). One example of exhumation of the Dwyka surface is to be found in north Shaba (Democratic Republic of Congo), where a pre–300 Ma topography with upstanding inselbergs is reported to have been partially exhumed from beneath the Permian Lukuga cover rocks (Dumont, 1991). Surfaces exhumed from beneath Karroo sediments have been widely recognized in southern Africa. King (1951) devoted an entire chapter to the pre-Karroo landscape of Africa. In many places, the Dwyka subglacial surface is now reoccupied by the African Surface (see, e.g., de Wit 1999, p. 734). Some landforms can be extremely ancient: in southeast Namibia, older Precambrian granitic inselbergs have been exhumed from beneath the protective cover of late Precambrian Nama sandstones (Lageat, 1989a). In Tanzania, the inselberg landscape reported by Bornhardt (1900) is essentially pre-Cretaceous and exhumed from beneath overlying Cretaceous marine strata (Willis, 1936). These occurrences are far from representing an exhaustive list, but their existence highlights the palimpsest that is responsible for the polygenetic landscape of the African continent. In that sense it shares many similarities with the Australian landscape, where Cretaceous strata stripped during the Cenozoic have re-exposed upstanding granitic inselbergs and exhumed their surrounding pediments from beneath unconformities (Twidale, 1997). Likewise, valley-and-ridge topography of the eastern Kimberley region of northwestern Australia represents structurally controlled topography exhumed from more recently stripped Proterozoic rocks (Young, 1992).

Lithologically controlled escarpments were probably among the most important of surface irregularities on the African Surface. Gently dipping sandstone strata of various geological ages lying unconformably on the basement, as is also the case in South America and Australia, play a major role in maintaining crowns of relief above the main plateau surfaces. In West Africa, features of that kind are well developed at the edges of the outcrops of gently dipping late Precambrian and Paleozoic sandstones, including the Taoudeni and Bove sandstones. Dolerite sills, many of which were erupted ca. 200 Ma during the CAMP (Central Atlantic Magmatic Province) plume episode (e.g., Marzoli et al., 1999) have generated similar, <100 m steps in the African Surface topography of West Africa. In central Africa, the Wadda and Gadzi sandstone plateaus are partly autonomous land surfaces, highly degraded into pseudo-karstic features that have been in part geomorphologically separate from the evolution of the lower plateaus. That is strong evidence of the preservation of irregularities on the African Surface that date from the time during which the surface was actively being eroded. Compositional irregularities among highly deformed rocks in the Birrimian greenstone belts of West Africa have also been recognized to have influenced local relief on the African Surface (Gunnell, 2003). In SE Africa, Lesotho and the Lyndenburg Drakensberg stand as giant buttes on the edge of the plateau, and overlook both the coastal plain and the continental hinterland. This suggests that resistant islands of relief existed throughout the Cenozoic and would have risen piggyback on the swell flanks.

Another important control on the elevation of the African Surface has been emphasized by Şengör (2001), who pointed out that because the intracontinental rifts (he considered the Anza rift) of Afro-Arabia were receiving sedimentary deposits from rivers flowing along the rifts over much of the time during which the African Surface was evolving, there is likely to have been relief on the rift shoulders. Using AFT (apatite fission tracks), Foster and Gleadow (1992, 1996) reached a similar conclusion for the Anza rift during the 65–40 Ma interval. Those rift shoulders could have been subject to renewed elevation whenever they were subjected to stress change in the continental crust. Rift shoulders may therefore have locally represented the greatest positive topographic anomalies on the African Surface. Sedimentary infill in the rifts indicates that rift shoulders are likely to have reached as much as 0.5 km of elevation over the regional African Surface pediments even during the interval of tectonic quiescence. Rift shoulders are thought to have played another role in being eroded to supply sediment in the direction away from the rifts to generate thin but widespread sedimentary deposits, such as the Cretaceous "Nubian" sandstone, on the African Surface in the continental interior. The "Nubian" sandstone was deposited on the African Surface in an extensive region, most likely in megafans (Wilkinson, 2004), on the south side of the area presently occupied by the Sahara, at a time when the Central African, Sirt, and Abu Gharadig rifts (see Fig. 6) were active. The shoulders of all those rifts could have provided sources of sediment (Badalini et al., 2002).

Igneous activity is also likely to have been associated with locally rugged topography across the African Surface. During the interval of tectonic quiescence, many areas of igneous activity are likely to have been associated with higher ground, including kimberlites in southern Africa (Hawthorn, 1975; Lageat, 1989b), ca. 50 Ma alkaline intrusions of Namaqualand, the 65–30 Ma granites of Cameroon, many igneous bodies in Egypt (Guiraud and Bosworth, 1999), and the basalts of southern Ethiopia (45–35 Ma).

This has not been an attempt to review exhaustively possible sources of irregular surface elevation on the African Surface. The aim has been rather to draw attention to some of the main influences likely to have contributed to those irregularities, which were all small relative to the size of the African continent.

The African Surface in Basins That Experienced Marine Incursions between 100 Ma and 34 Ma

It is clearly in places where basins experienced marine incursions lapping onto the African Surface, or onto older surfaces from which sedimentary rocks had been beveled by the African Surface, that the local chronology of geomorphic events can be best reconstructed. Global sea level was generally high during Late Cretaceous times (100–65 Ma) and much of Africa was sometimes flooded. Marine and nonmarine (Mateer et al., 1992) Cretaceous sedimentary rocks were deposited in intracontinental rifts in episodes that occurred during a protracted interval between the Cenomanian (ca. 95 Ma) and the early Eocene (ca. 50 Ma; Guiraud et al., 1992) in the Ténéré (Niger), Muglad (western Sudan), Benue (Nigeria), Sirt (Libya), and Anza (Kenya) Basins. Most of these rift basins experienced marine incursions, although none are currently known in any of the Sudan rifts. Sea level began to subside during the Paleocene (from 65 Ma) and after a brief recovery resumed subsidence at the end of Paleocene times (ca. 55 Ma; Haq et al., 1987, Zachos et al., 2001). As a result, while the African Surface was evolving, the coasts of Afro-Arabia were repeatedly flooded during highstands, with the greatest transgressions reaching far into the continent. Sahagian (1988) used Cenomanian transgressive surfaces to estimate magnitudes of post–100 Ma surface uplift across Afro-Arabia (Fig. 8A). He obtained fairly good agreement with Bond (1978, 1979), who had used hypsometric curves for the similar purpose of demonstrating late Cenozoic uplift of Africa relative to North America, South America, Australia and Europe. Bond was criticized by Harrison et al. (1983) for his choice of time slices and for some statistical arbitrariness. Sahagian (1988), in a conclusion that is very similar to that of this paper but which is based on a very different and highly specific method of analysis, suggested that many of the Cenozoic swells of Africa involved the upwarping by ~3 km of land surfaces that were initially flat and had been close to sea level at 100 Ma. By using a Cenomanian depositional datum plane, Sahagian effectively addressed the deformation of older strata that occurred during the past 30 m.y. Evidence from South Africa shows that the late Campanian–early Maastrichtian transgression left carbonate rocks preserved as small outliers that have been elevated during the past 30 m.y. and today stand as high as 367 m a.s.l. (at Needs Camp; see Reyment and Dingle, 1987). Approximately one-sixth of this magnitude of post-Cretaceous surface uplift was achieved recently, because

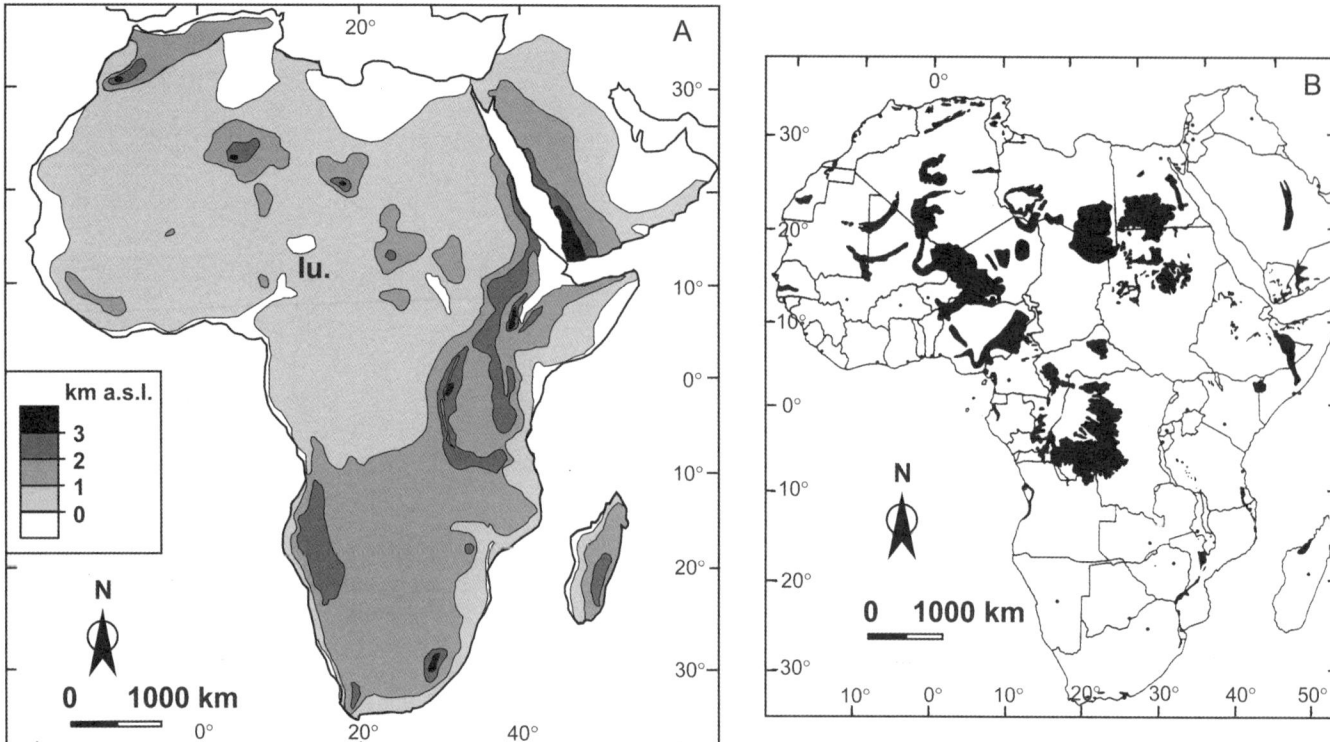

Figure 8. Evidence of upward warping of the African Surface gauged from marine (A) and nonmarine (B) sediment distribution. (A) Mean Cenomanian (ca. 90 Ma) topographic surface corrected for sediment loading by post-Cenomanian sedimentary rocks, simplified after Sahagian (1988). Areas of basin inversion and topographic uplift are reconstructed using marine Cenomanian rock distribution. The approximate Cenomanian paleoshoreline (white areas below 0 km) is indicated for reference, revealing that ca. 90 Ma little of Afro-Arabia was submerged except in the north, and that the Iullemeden Basin (Iu.) lay within the trans-Saharan seaway. An added correction for a ~0.25 km decrease in sea level since Cenomanian times would provide an even more striking picture of post-Cenomanian tectonic deformation, which occurred as swell uplift (and local rift-flank uplift in East Africa) in post-Eocene times. (B) Distribution of nonmarine Cretaceous outcrops in Afro-Arabia (smaller outcrops excluded), after Mateer et al. (1992). These rocks are usually Lower to Middle Cretaceous in age, and stratigraphically older than the Cenomanian. Upper Cretaceous nonmarine strata also occur extensively, particularly from Mali to east Niger (Coniacian to Maastrichtian, termed *continental hamadien*), in the Benue trough and Chad Basin (Gombe sandstone), and in the Congo Basin. Although these rocks were not necessarily deposited close to Cretaceous shorelines, they all relate to rift-shoulder erosion subsequent to the Santonian rifting event. As such, these nonmarine deposits define the relative location of eroding topographic highs and lower-lying sediment sinks during Cretaceous times. A survey of their current elevations would thus also contribute to an understanding of the post-Cretaceous topographic deformation of Africa.

Quaternary marine terraces rise to over 60 m along the KwaZulu-Natal coast (Orme, 2005). Farther north, in Mozambique, the timing of uplift is equally supported by the observation that shallow-water marine sediments deposited on the wide continental paleoshelf during the Miocene presently outcrop on the coastal plain (Salman and Abdula, 1995). The approach of Sahagian was limited to the extent that he did not consider the nonmarine Cretaceous sedimentary rocks that are abundant in many regions of Africa (see Mateer et al., 1992) although not necessarily deposited close to Cretaceous shoreline elevations (Fig. 8B). Here consideration is given to the African Surface as a land surface that developed over a long time span encompassing the Cenomanian, and involving the deposition of some nonmarine Cretaceous sediments that were later also uplifted.

The Iullemeden Basin (Greigert, 1966; Kogbe, 1981; Moody and Sutcliffe, 1991), situated roughly 1000 km from the Gulf of Guinea in the interior of Africa, provides evidence of how the sea transgressed episodically onto the evolving African Surface in a trans-Saharan seaway linking Algeria to the Gulf of Guinea perhaps through the Benue trough. Marine incursions occurred during Turonian time (ca. 92 Ma) as well as during Maastrichtian to Paleocene (ca. 75–60 Ma) times. Marine and nonmarine sedimentary rocks up to 625 m in thickness accumulated at an average rate of 10 m × m.y.$^{-1}$ in the Iullemeden Basin between 100 Ma (the end of Albian times) and 34 Ma (i.e., the end of Eocene times; see Boudouresque et al., 1982). The unusual thickness of sedimentary rocks deposited on the African Surface in the Iullemeden Basin is attributable to location of the basin over the Gourma rift of Proterozoic age (Kogbe, 1981). Five marine transgressions and four episodes of the development of surfaces characterized by surface alteration that led to the formation of iron-oxide minerals, including Late Cretaceous and Paleocene oolitic bauxite units, have been described from the Iullemeden Basin (Fig. 9). Over the large areas of the Afro-Arabian continent, on which no record of marine transgressions between the

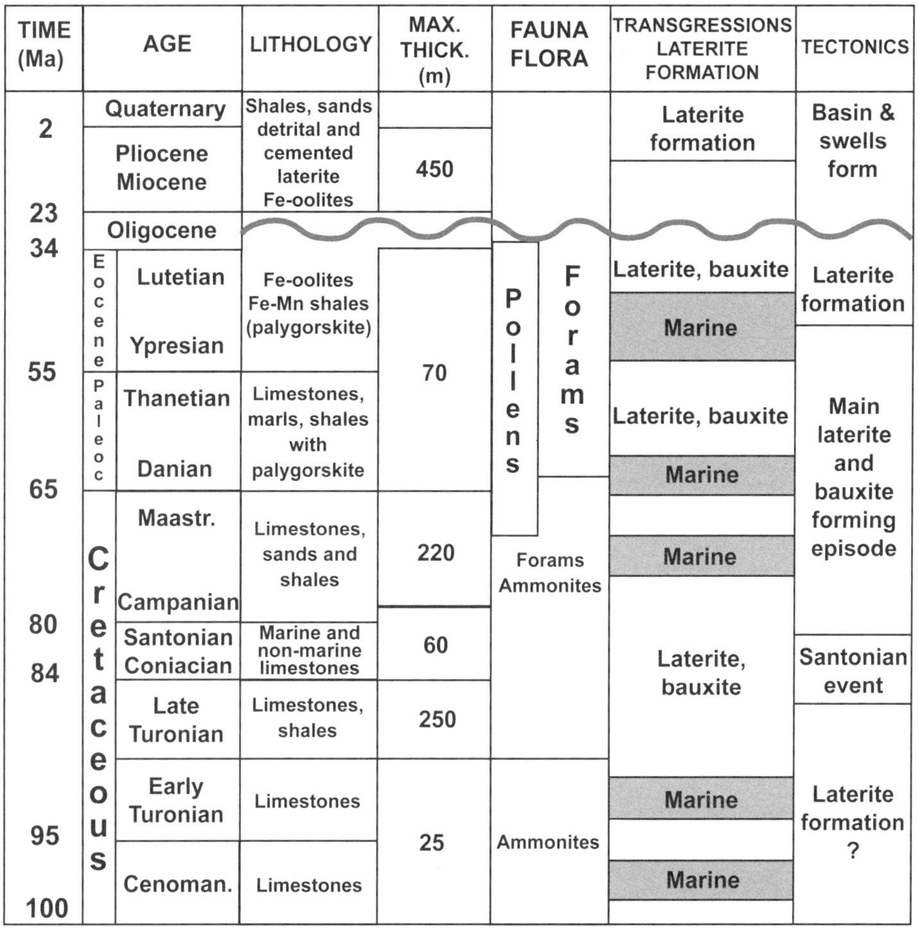

TIME (Ma)	AGE		LITHOLOGY	MAX. THICK. (m)	FAUNA FLORA	TRANSGRESSIONS LATERITE FORMATION	TECTONICS
2	Quaternary		Shales, sands detrital and cemented laterite Fe-oolites	450		Laterite formation	Basin & swells form
23	Pliocene Miocene						
34	Oligocene						
55	Eocene	Lutetian	Fe-oolites Fe-Mn shales (palygorskite)	70	Pollens / Forams	Laterite, bauxite	Laterite formation
		Ypresian				Marine	
	Paleoc	Thanetian	Limestones, marls, shales with palygorskite			Laterite, bauxite	Main laterite and bauxite forming episode
65		Danian				Marine	
	Cretaceous	Maastr.	Limestones, sands and shales	220	Forams Ammonites	Marine	
80		Campanian					
84		Santonian Coniacian	Marine and non-marine limestones	60		Laterite, bauxite	Santonian event
		Late Turonian	Limestones, shales	250			
		Early Turonian	Limestones	25	Ammonites	Marine	Laterite formation ?
95							
100		Cenoman.	Limestones			Marine	

Figure 9. Laterite formation and marine transgressions on the African Surface between 100 and 30 Ma reconstructed from the stratigraphy and paleogeography of the Iullemeden Basin in the interior of West Africa. After Boudouresque et al. (1982), who identified five transgressive episodes each between ca. 100 Ma and ca. 40 Ma, and followed by an unconformity and subsequent accumulation of the *continental terminal*, here shown with a maximum thickness of 450 m of Pliocene and Miocene sedimentary rocks.

end of Albian times and the end of the Eocene (100–34 Ma) has been preserved, the composite African Surface record (corresponding to that which can be subdivided within the Iullemeden Basin) locally includes some thin nonmarine sedimentary rocks: for example units of the mainly Cretaceous *continental intercalaire* and *hamadien*. The record, however, is dominantly of erosion and deep weathering. The Iullemeden Basin, and perhaps also the Congo Basin, where Dumont (1991) analyzed the succession of erosional cycles in Shaba province using the Cretaceous nonmarine Kwango siliciclastic deposits, may be among the best places in which to characterize the composite nature of the African Surface in interior Africa.

Marine intercalations and locally iron-enriched cappings sealing erosion surfaces have been recognized in other sedimentary basins, mostly at Afro-Arabia's continental margins (Figs. 10 and 11). In those places, the composite African Surface can be resolved into local separate surfaces. In the following section, more local aspects of the evolution of the African Surface are emphasized in a regional survey of erosion surfaces in Afro-Arabia. Attempts are made to place the African Surface both within the present landscape and within the suite of surfaces that have been previously reported.

REGIONAL CHARACTERISTICS OF THE AFRICAN SURFACE

The preservation, extent, and character of the African Surface vary greatly, and research has emphasized different aspects of the surface in different parts of the continent. For those reasons, we discuss the African Surface in separate areas, noting from the outset that much more is known in some areas than in others.

The African Surface in Its Type Area: Southern Africa

History of the Concept of the African Surface

The term "African Surface" in its type area has a long and complex history. The work of L.C. King, who made distinctive contributions to the study of erosional surfaces in Africa and indeed in the entire world during several decades in the mid–twentieth century, provides a good starting point. Various erosional surfaces in South Africa, including the African Surface, were defined and mapped by King. Unfortunately a great deal of confusion has arisen from the fact that King himself repeatedly revised either his own definition of the African Surface or the names that he gave to various surfaces from one publication

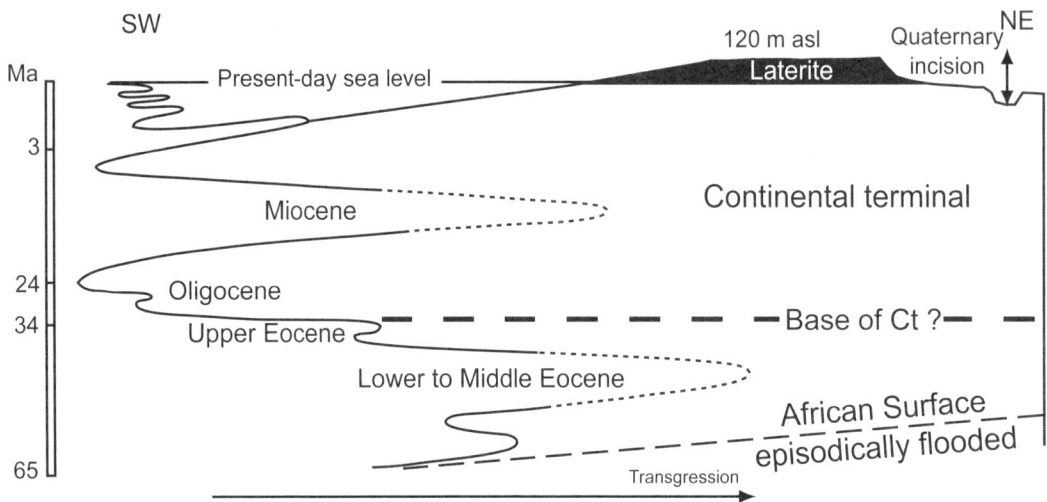

Figure 10. Marine intercalations (left side of figure) within sedimentary rocks deposited on the African Surface in Senegal and Mauritania during Cenozoic times (modified after Kogbe, 1981). Note that in this figure, as in some other work, the *continental terminal* (Ct) is shown to reach down into the Eocene. The Ct is not a formal lithostratigraphic unit, or series, with a type locality or defined age range. This probably means it will have to be abandoned. It may be preferable in future to restrict the Ct nomenclature to terrigenous rocks generated after 30 Ma as a consequence of swell uplift because, as in many other areas of the Tropics, the "Eocene" ages of the floras and faunas of such Cenozoic red beds are poorly established. The African Surface was episodically flooded, and the major Oligocene regression, which coincides with the establishment of the Antarctic ice sheet as well as the beginning of the rise of Africa's swells, is revealed in much of Africa's offshore sedimentary and onshore geomorphic record by the occurrence of Ct red beds in coastal catchments. Erosion continued during subsequent swell uplift. Horizontal extent of the regions illustrated is typically tens of kilometers.

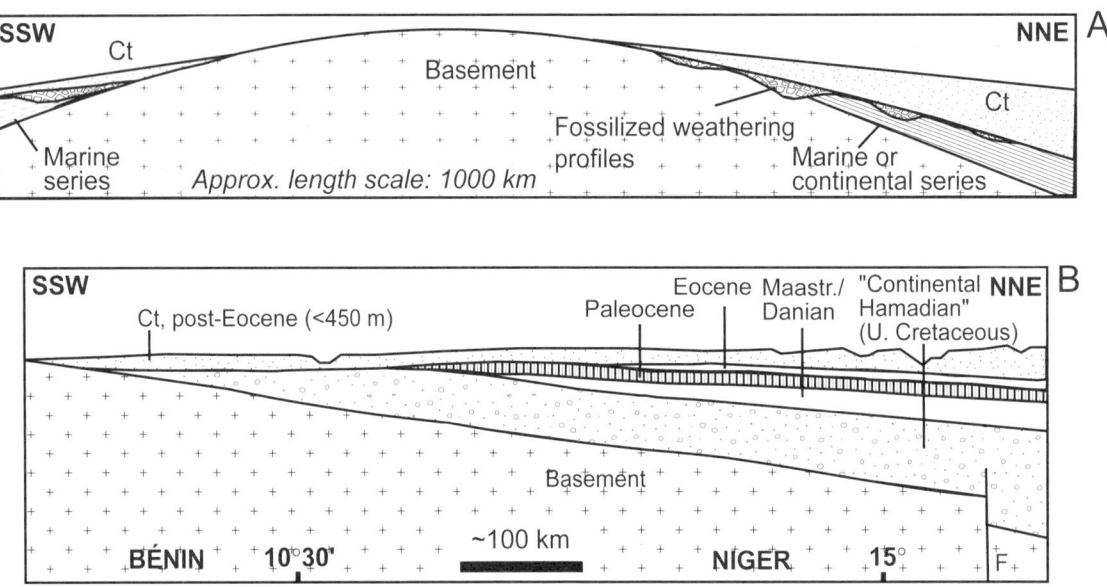

Figure 11. Two schematic profiles showing the character of the African Surface in the region north of the Gulf of Guinea and close to the prime meridian. Cross sections located on Figure 2. (A) Schematic north-south section across the West African swell, which began to be elevated ca. 30 Ma, showing asymmetry of inner and outer swell-flank basins. To the south, marine intercalations occur within sedimentary rocks deposited on the African Surface in the Gulf of Guinea during the Cenozoic. Marine sedimentary rocks in the interior are pre-Cenozoic in age, having been deposited in the trans-Saharan Cretaceous seaway. The sedimentary rocks deposited on the African Surface in the northern part of the profile record a complex series of marine transgressions and regressions during which subaerial weathering has generated paleosols. The *continental terminal* (Ct) has been deposited on the flanks of the swell as it has risen. (B) Profile corresponding to the northern part of the top profile from the swell into the Iullemeden Basin. The *continental hamadien*, spelled Hamadian in figure, represents part of the *continental intercalaire*, which is a name given to nonmarine Cretaceous sedimentary rocks deposited on the African Surface in West Africa. The *continental terminal* (Ct) has been deposited on the flanks of the swell and in the Iullemeden Basin since the end of Eocene time (34 Ma) as the swell has risen. After Lang et al. (1990).

to the next without clearly discussing or explaining his motives (see Fig. 12 and cf. King 1948, 1951, 1963, 1976). In the second (1951) edition of his *South African Scenery*, King attributed a fundamental role to the Gondwana surface as being the ancestral peneplain of Africa, which had developed between the Early Jurassic and the Early Cretaceous. King also considered the Gondwana surfaces landward and seaward of the Great Escarpment as "probably without rival for smoothness of beveling anywhere in the world (p. 246)." By the third (1963) edition, the map of his erosion surfaces exhibited a significant proportion of remapping of his former Gondwana surface under the appellation "African Surface." It was now the African cycle that came to represent what King called "planation *par excellence*" (p. 205–209), as such being "the most readily recognized of all the various cyclic land surfaces" (1963, p. 208). In 1951, King had considered that, as Gondwana broke up at the Jurassic-Cretaceous boundary, a new erosion cycle was introduced not by continental uplift, but by "betrunking of river systems and roughing out of the African coastline" (p. 246–250): this set the stage for what he then called the African cycle, which generated the African Surface by the end of the Paleogene. By 1963, King had relabeled the African cycle as "post-Gondwana cycle." It dissected the Gondwana land surface into broad valleys and plains. This post-Gondwana topography was therefore the "first landscape cycle of Continental Africa." The post-Gondwana surface of 1951 thus continued to represent the land surface developed in response to the breakup of Gondwana during Early and Middle Cretaceous times, but the African Surface was defined as being the terminal result of protracted denudation initiated by the breakup of Gondwana. Topographic smoothing increased with time. In this new definition, the African Surface therefore acquired the morphologic attributes originally recognized in the Gondwana surface of 1951, thereby effecting not only significant semantic drift in the classification scheme, but also a radical change in the understanding of the landscape cycles of southern Africa.

In representing the ultimate planation of the post-Gondwana cycle, the African Surface constitutes the fundamental topographic benchmark of the African landscape. This late stage of post-Gondwana relaxation was in progress between the Late Cretaceous and the late Eocene. Mid-Cenozoic (post–ca. 30 Ma) deposits locally lie on top of the African Surface so that it has widely been assigned a mid-Cenozoic terminal age. It has also been termed by some workers the "mid-Tertiary" surface (e.g., Dixey, 1956). Large areas of what had been labeled by King in 1951 as the "African Surface" were annexed to a newly introduced "post-African Surface," attributed to rejuvenation caused by a resumption of crustal movements during Neogene times. A comparison of the geomorphic maps from *Morphology of the Earth* (1962, Fig. 119 therein) and the 1951 edition of *South African Scenery* clearly reveal this puzzling discrepancy (Fig. 12). King's apparent hesitation over the relative age and geographic labeling of surfaces also affected his interpretations in Uganda, another type area of African geomorphology. King at one time suggested that the Buganda surface was his Gondwana surface. He subsequently changed that attribution to make it his African Surface.

King (1976) later decided to further relabel his South African surfaces, without, however, changing their age assignments as he had between 1951 and 1962. While the Gondwana surface retained its name, the post-Gondwana surface became the Kretacic surface, the African Surface became the Moorland surface, the post–African I surface became the Rolling surface, the (Pliocene) post–African II epicycle became the Widespread surface, and the Youngest Cycle described a Quaternary cycle of rejuvenation. None of these terms have, however, gained much currency, even within South Africa.

Ollier and Marker (1985) revisited King's African Surface type area and returned with a very different reading of the landscape. Where King (1976) had distinguished the five or six erosional levels listed above on the Highveld Plateau of South Africa, Ollier and Marker defined just two surfaces: an upper "paleoplain" at 1.5 km a.s.l. and the coastal plain. The paleoplain was considered to be a composite patchwork of Gondwana surface relics and African Surface tracts. It was suggested to be little more than a modified version of the land surface that existed before the breakup of Gondwana. By 1985, although nobody denied the reality of the African Surface, controversy thus persisted over its extent, distribution, age, and characteristics even in the historic type area of African geomorphology that was South Africa.

Recent Interpretations of the African Surface

A comprehensive reinterpretation of South African erosion surfaces by Partridge and Maud (1987) represents the beginning of the modern era of the study of the African Surface in southern Africa. That study contained two important innovative aspects: (1) The conclusion that no surfaces can be identified on the African continent that are older than the breakup of Gondwana. Subsequent workers have generally concurred with that conclusion, the recognized exceptions being the ancient erosion surfaces that have been exhumed from beneath overlying sedimentary rock accumulations (see earlier section). (2) A strong emphasis on the elevation of what in that paper is called the "Great South African Swell" during the past 20 m.y. That interval corresponds, within present resolution, to the interval identified as post–30 Ma in this paper.

Partridge (1997) followed what was becoming a widespread practice by using the term "African Surface," but they used that term in a new way. For them, the African Surface was a surface that had formed as a result of complex cycles of erosion between the time of a Gondwana rifting event, which they put at 150 Ma, and the early Miocene ca. 20 Ma. Their definition of the African Surface closely resembles the definition of the African Surface used in this paper. There is, however, one distinctive property of Partridge and Maud's African Surface that is not compatible with the way in which the term African Surface is used here. Partridge and Maud (1987) applied the term both to a composite surface above the Great Escarpment of Southern Africa and to a surface that extends toward the coast from the foot of that typically 1-km-high erosional escarpment.

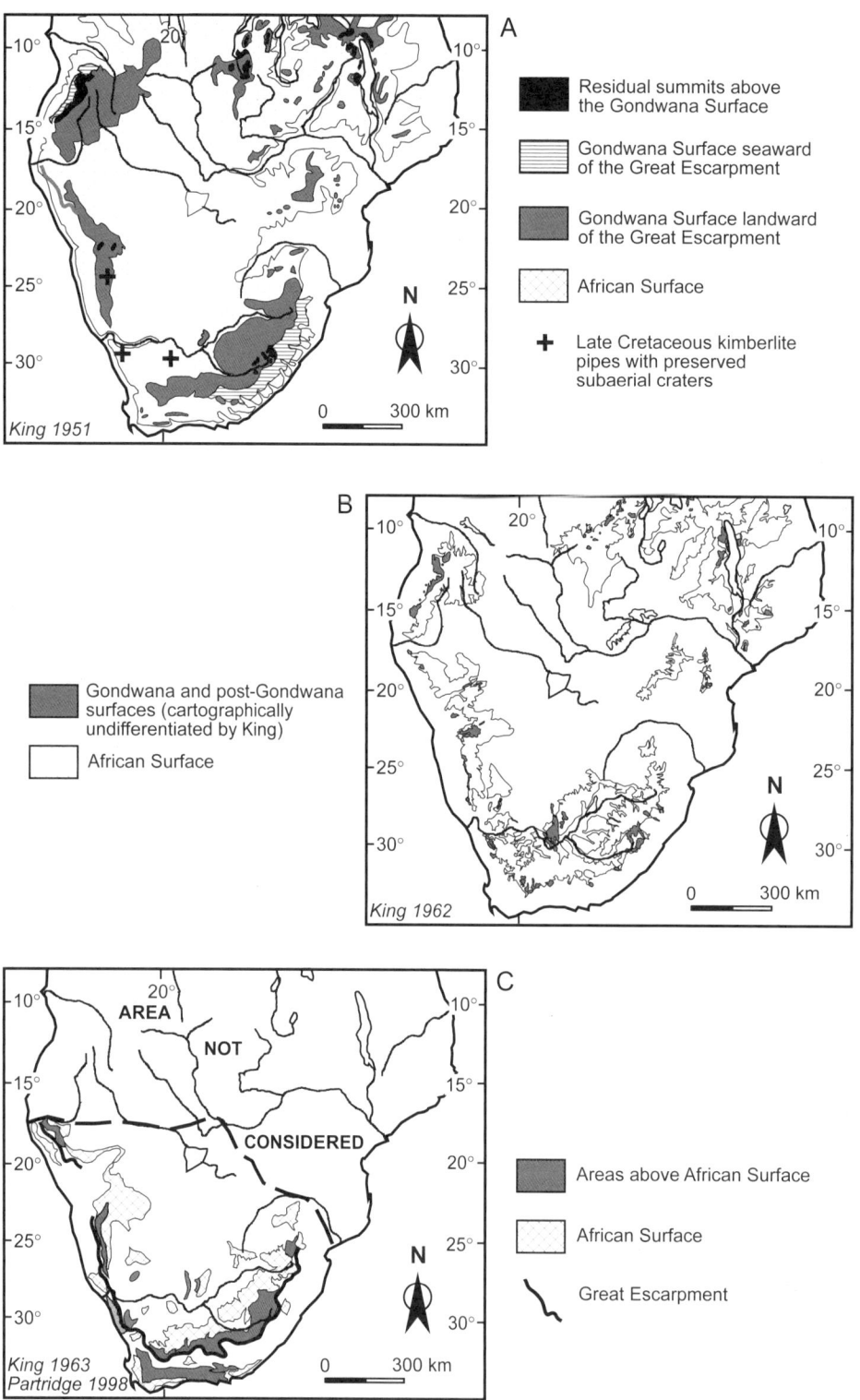

Figure 12. History of African Surface recognition and reconstruction in its type area, southern Africa. Distribution of erosion surfaces is after (A) King (1951), simplified; (B) King (1962, Fig. 119 therein); and (C) King (1963, p. 202–213) and Partridge (1998). Figure 12C mostly reflects minor readjustments to Figure 12B, but involves greater consideration for the morphology of terrain seaward of the Great Escarpment (e.g., Cape Fold Belt). Late Cretaceous kimberlite pipes with preserved primary subaerial craters (after Hawthorn, 1975; Cornelissen and Verwoerd, 1975; Janse, 1975; Smith, 1986; Mitchell, 1986) reveal that depths of post–African Surface aerial denudation in those areas as on many summits of the higher Drakensberg on the basis of Karoo stratigraphy (e.g., Eales et al., 1984) have been small in spite of localized dissection by fluvial activity. This is in broad agreement with apatite fission-track data, which detect islands of low erosion close to those areas (see Fig. 32).

They further distinguished two younger erosion surfaces that cut into the surface below the Great Escarpment (see Fig. 12C), giving those surfaces the names post–African 1 surface and post–African 2 surface. King (1962) had also used those terms, but his meanings were not quite identical to theirs.

Most appear now to agree that the African Surface has been uplifted and is presently higher above sea level than it was during most of its erosional period. Partridge and Maud (1987) concluded that there had been two phases of uplift during the past 20 m.y. The first phase involved 150–300 m elevation during the early Miocene, i.e., soon after 22.5 Ma; and the second involved an additional increment of 900 m since ca. 5 Ma. It is not clear what these uplifts achieved in terms of modifying what Partridge and Maud (1987) apparently considered to be an already existing, and generally 1-km-high, Great Escarpment.

From a study of Turonian (ca. 90 Ma) kimberlite pipes in the Kimberley area, and from the model of kimberlite pipe emplacement of Hawthorn (1975), Lageat (1989b) concluded that a thickness of ~1.4 km of Karroo sediments had been eroded since Turonian times (ca. 92 Ma) from the surface of the former Orange Free State and northern Cape Province. This estimate is consistent with the AFT denudation results of Gallagher et al. (1998) and Brown et al. (2002). Lageat tentatively related the 1.4 km of erosion to a phase of deformation that he correlated with the "Upper Cretaceous epeirogenic phase" of Dingle et al. (1983), which involved massive outwash of terrigenous sediment occurring between the Turonian and the Santonian (ca. 94–84 Ma). That erosion supplied the Orange River drainage and formed prograding fan deltas covering the horst and graben structures of the Atlantic margin of South Africa. Jungslager (1999, particularly Figs. 2 and 14 therein) provided more up-to-date information about offshore depositional history close to the Orange River mouth. This is generally compatible with Dingle's older analysis, although the prograding fan deltas seem to have developed over a longer interval than Dingle suggested. An Afro-Arabian-wide tectonic episode that occurred ca. 84 Ma during Santonian time has been attributed by Guiraud and Bosworth (1997) to arc collision on the Tethyan shore of Arabia.

Rapid erosion occurring in the Kimberley region of South Africa may have been associated with that event (see also Burke et al., 2003a, p. 46–47). In support of the idea of a phase of tectonic deformation during the Late Cretaceous, Lageat (1989b, 1997) linked the sculpting of structurally controlled homoclinal landforms of the eastern Bushveld Complex to stages in crustal deformation and landscape development. Lageat showed that on the eastern limb of the Bushveld Complex, the landscape is characterized by the rejuvenation of the Highveld surface, which is the name applied to the African Surface in that region. Elevated remnants of the African Surface that are well preserved on the felsic roof-rocks of the Bushveld Complex developed, he suggested, in the wake of a Turonian (ca. 91 Ma) phase of tectonism and related denudation. Mafic intrusive rocks of the same area, which were less resistant to weathering and stripping than were the felsites, form topographic lows. In the northwestern outcrop area of the Bushveld Complex, by contrast, the felsic roof-rocks and the mafic intrusives are all beveled by the African Surface, here locally termed the "Bushveld surface." Instead of being differentially eroded into a suite of structural landforms, the topography in this northwestern outcrop remains unaffected by the lithologic variability of the complex. The African Surface slopes gently toward the Kalahari Basin and exhibits only a few, more resistant, gabbro inselbergs. Lageat (1989b) attributed the morphologic contrasts between the eastern and western limbs of the Bushveld Complex, i.e., subdued relief in the west and bold homoclinal landforms in the east, to nonuniform uplift. This geomorphic asymmetry of the otherwise fairly symmetrical, saucer-shaped Bushveld Complex was attributed to deformation of the African Surface, involving asymmetric updoming. The eastern rim of the Bushveld Complex was affected by the uplift of the Great Escarpment between the Lebombo and the Drakensberg, so that the western rim is lower as a result of the upward flexure of the eastern rim. The southern rim of the Bushveld coincides with the divide between the Orange and Limpopo drainage systems and the northern rim is affected by vertical displacements associated with a major fault system, the Palala Shear Zone.

Similar interpretations of African Surface history in this area had been discussed by Du Toit (1933), and more recently by Stratten (1979) and Mayer (1985), who showed that some of the diamondiferous alluvial gravels currently at an altitude of 1.5 km a.s.l. in the North West Province (Lichtenburg area, formerly southwestern Transvaal) were derived from a source region that now lies at only 1.1 km.

Although Lageat (1989b) emphasized Late Cretaceous deformation and related erosion, he also considered that the Highveld (alias African) surface began to be dissected again in Mpumalanga Province (formerly eastern Transvaal) during Oligocene times (34–22 Ma) as a consequence of renewed uplift. He associated that uplift with increased terrigenous input to the NE Natal margin, where Oligocene rates of siliciclastic influx rose to nearly 400% of their early Cenozoic rates at the mouth of the Limpopo. Burke (1996, p. 396) made a similar point by showing that although both the Limpopo and the Zambezi Rivers had generated deltas during the Late Jurassic and Cretaceous (ca. 160–65 Ma) no Limpopo and Zambezi deltas existed during the earlier Cenozoic (ca. 65–30 Ma; Salman and Abdula, 1995). The present deltas, like that of the Tugela River (Fig. 9) began to form ca. 30 Ma. According to Lageat's (1989b) analysis, 34–30 Ma would have been the terminal age of the African Surface in northeastern South Africa. Lageat's conclusions also indicate that the Limpopo/Mpumalanga and Natal Drakensberg escarpments are relatively recent morphologic features, or at least have acquired a large proportion of their relief since the Oligocene. Lageat (1989b) recognized an early Miocene (22–16 Ma) age for a Middleveld surface that affected the terrain seaward of the Natal and Limpopo/Mpumalanga Drakensberg. In the late Neogene (since 6 Ma), a Lowveld surface developed. Partridge and Maud (1987) had identified two similar surfaces in this area.

Burke (1996) addressed the erosion of southern Africa as part of a general analysis of the erosion of Africa. He also, rashly, used the term "African Surface" without considering the many different ways in which it had already been used. Partridge (1998) and Partridge and Maud (2000) more recently revisited the topic of the erosion surfaces of South Africa. Those treatments differ only slightly from the fuller discussion of Partridge and Maud (1987). Discussions of additional topics, including apatite fission-track interpretations and mantle circulation in relation to continental elevation, are included. De Wit (1999) recognized a region of exposure of remnants of the African Surface distributed over an area in excess of 100,000 km² south of the Orange River, around 30°S, 20°E. The region lies at elevations between 900 and 1200 m a.s.l., and AFT estimates (Brown et al., 1990) indicated that most of the erosion of the surface took place during the Cretaceous. The preservation of crater facies of Late Cretaceous and early Cenozoic alkaline intrusions was interpreted as indicating a substantial decline in erosion rates at the end of the Cretaceous that persisted into Cenozoic times. Extensive calcrete on the African Surface, which is locally overlain by a thin sheet of coarse red sand, was correlated with regionally developed late Miocene calcretes (Ward and Corbett, 1990). De Wit (1999, p. 724) concluded that "the geomorphology … has not evolved significantly since [Late Cretaceous] time and only minor erosion and incision has taken place [on the African Surface] in the last 60 [m.y.]."

In fitting his local study into the regional geomorphological development of southern Africa, de Wit (1999, Fig. 20A therein) mapped a Kalahari River (an ancestral Orange River) and farther south a Karroo River as flowing during Cretaceous times on the low-relief African Surface (Fig. 13). Between them the two rivers drained an area that covered the western half of South Africa and exceeded 0.5×10^6 km² in area. The valleys of the two rivers were broad, as is the valley of the Orange River

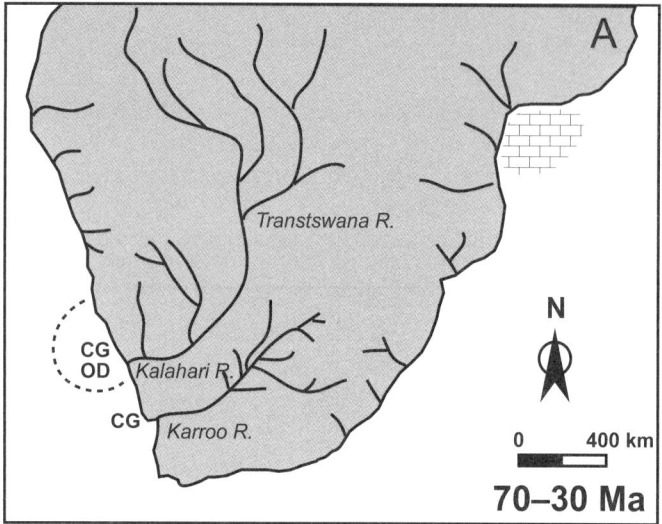

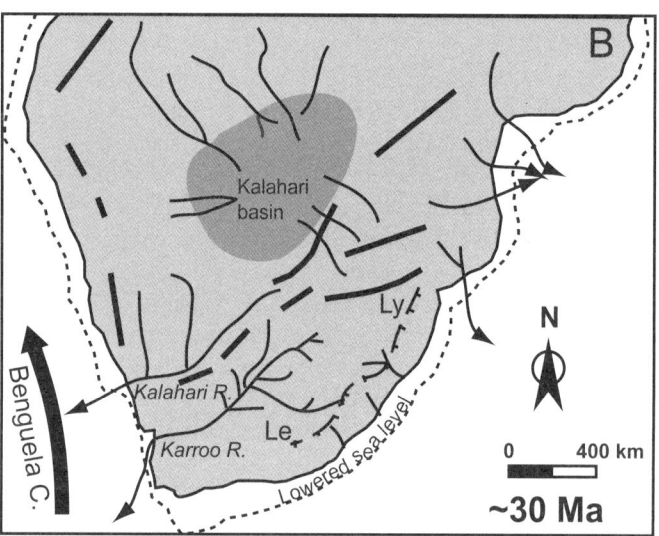

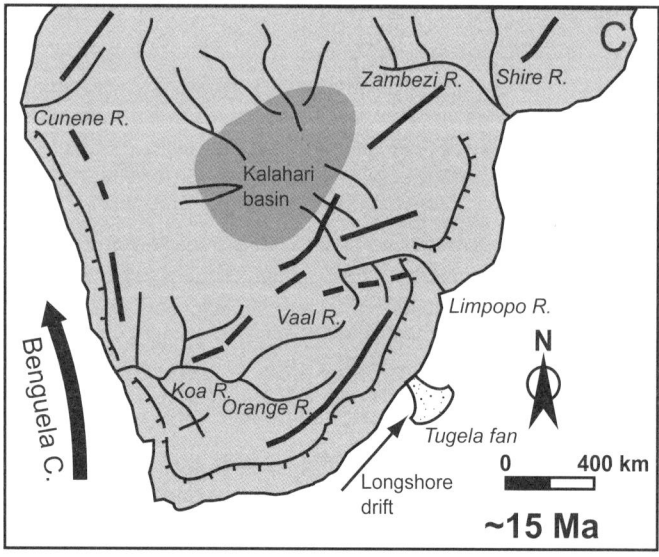

Figure 13. A revised later history of the African Surface in its type area (southern Africa) based on recent data. The African Surface occupies the area above the Great Escarpment (line with coastward facing ticks). A 30-Ma and younger surface (or a flight of two surfaces [Lageat 1989b]) lies between the foot of the escarpment and the sea. Erosion along major rivers has dissected the African Surface far into the interior in some places. (A) The mature African Surface on a low-lying Africa. CG—areas of Cretaceous diamondiferous gravels; OD—region of maximum below-shelf-break deposition from the Kalahari (alias Orange) drainage during Turonian to Campanian times (90–80 Ma) (after Jungslager, 1999). (B) The African Surface beginning to be elevated on the Great Swell at 30 Ma. The African Surface began to be depressed in the Kalahari at this time. Sub-swell axes (thick black segments) are from Du Toit (1933). New submarine canyons are indicated by thin arrows, and the possibility of a lithologically controlled persistent escarpment, or major residual relief masses between the Lyndenburg (Ly) and Lesotho (Le) megabuttes, is shown with a discontinuous line. (C) The Drakensberg escarpment in retreat ca. 15 Ma. The new Zambezi, Limpopo, and Tugela deltas, and deep-sea fans are active at this time. Longshore drift diverted sediment from coastal streams northeastward. The Limpopo River has captured the Kalahari (Orange) drainage. Cape Fold hard rock residual not shown.

above the Aughrabies Falls today. Cretaceous diamond-bearing gravels that were deposited at three places within the more than 1000-km-long valley of the Karroo River are among the few as yet identified detrital sedimentary rocks of Cretaceous age deposited on the African Surface in southern Africa (Fig. 13A).

In southern Africa the Zambian, Zimbabwean, Namibian-Angolan, and South African swells (Fig. 2) all rise to ~1.5 km and only to greater elevations in places such as the Drakensberg, where they have been flexed upward in response to local intense erosion on one flank (Lageat, 1989b; Gilchrist and Summerfield, 1994). The general understanding that the high elevation of South Africa was acquired in post-Eocene times is at odds with the suggestion of Cox (1980, 1993) that magmatic underplating beneath the Kaapvaal craton as a consequence of Karroo plume-related eruptions 180 m.y. ago accreted a thickness of 2–6 km of gabbroic material to the base of the crust—such accretion providing South Africa with ~1 km of excess elevation with respect to surrounding regions. This view is doubtful for two reasons: (1) any such underplate-related swell would have been eroded probably very quickly and certainly by late Cenozoic times; and (2) recent extensive seismic surveys of the Kaapvaal craton have not resolved any such underplate beneath the Precambrian crust (James et al., 2001). The South African swell is geologically much more recent than Cox suggested and involves 30 Ma and younger uplift of the African Surface. Our interpretation is also at odds with the suggestion of Lithgow-Bertelloni and Silver (1998) that the topographic swell of South Africa is unusually elevated because it overlies the hottest and most voluminous part of the anomalous deep mantle close to the core-mantle boundary under Africa. That seismically slow volume, now called the African LLSVP (Garnero et al., 2007), has come to be considered compositionally distinct and stable rather than buoyant (Torsvik et al., 2006). Moreover, plumes are now recognized to have risen from the LLSVP margins but not from above the interior (Davaille et al., 2005, and Torsvik et al., 2006). The LLSVP appears to have been situated where it is now for ~200 m.y. because large igneous provinces have been erupted at the surface from above the edge of that volume during those 200 m.y. (see Burke and Torsvik, 2004; and Burke et al., 2003a, Fig. 2 therein), whereas uplift of the Great Swell of South Africa only started ca. 30 Ma.

The African Surface in Southern Africa: A Summary

Although the term "African Surface" has been used in many ways in southern Africa, increasing similarities in the use of the term, if not as yet a consensus, are emerging from the work of recent years. That part of the African Surface of Partridge and Maud (1987) that extends landward of the Great Escarpment is the same composite surface as the African Surface of Burke (1996), and so is the African Surface of de Wit (1999). The "Upper Paleoplain" at 1.5 km of Ollier and Marker (1985) is also comparable because it is suggested to be a composite surface, i.e., a patchwork involving pre-breakup relics and African Surface tracts. Ollier and Marker's conclusion that "the Upper Paleoplain represents little more than a modified version of the pre-Gondwanan break-up surface" (Ollier and Marker, 1985) distinguishes their interpretation from those of Partridge and Maud and Burke, who avoided consideration of pre-Gondwana breakup surfaces. Lageat (1989b) also disagreed with Ollier and Marker about the preservation of residual pieces of pre-breakup surfaces. His "Highveld" plain is comparable to the African Surface and to the Upper Paleoplain of the three other authors, even though he regards the surface as having been initiated later. Because Ollier and Marker, Partridge and Maud, De Wit, and Burke all regard the African Surface as composite they would appear to be able to assimilate Lageat's Highveld surface as an element of their composite surface. That is particularly true if Lageat's estimate of the amount of erosion from the top of kimberlite volcanoes in the Kimberley area is high.

The composite African Surfaces of all the aforementioned authors combine the "post-Gondwana cycle" surface as elements of King's various models within a single composite African Surface. There remains some minor difference of opinion among these various authors as to when exactly the African Surface cycle ended, but there is agreement that the African Surface began to be dissected and deformed in response to new crustal movements. Authors converge toward a late Eocene (Lageat) to early Miocene (Partridge and Maud), i.e., 36–20 Ma, bracket for the initiation of deformation of the African Surface. Because of our Africa-wide and tectonically driven perspective in this paper, an age of ca. 30 Ma, i.e., a date that falls close to the middle of that 16 m.y. bracket, is here preferred. Given presently available data sets, resolution of the timing of the beginning of uplift of the Great Swell of southern Africa and of the start of erosion into the African Surface can hardly be better than 28 ± 8 m.y. Offshore deep-water results, as yet largely unavailable for South Africa, are likely eventually to provide better resolution. Whatever its precise timing, it was the beginning of deformation that terminated development of the African Surface in southern Africa. That development, in some areas, had lasted for a maximum of 150 m.y., although Lageat (1989b) opted for less than 75 m.y. in the Highveld Plateau area that he studied.

In summary, the evolution of terminology over the past 50 yr reveals (1) an unwavering acceptance of the African Surface as a geomorphic reality, (2) a considerable measure of geographic repositioning in terms of the extent and distribution of the African Surface, and (3) an increasingly noncommittal appreciation of the denudation chronology and partitioning of land surfaces above the African Surface. While initially also including the aggradational surfaces (e.g., Kalahari Basin) assumed to have been the sinks for the eroding upland source areas, the African Surface shrinks with time, but also replaces tracts of land initially labeled as Gondwana. Here we endorse King's African cycle of 1951 as the key event setting the stage for the African Surface as it developed over the next 150 m.y. However, we concur with the later revisions of King and Partridge in which the former Gondwana tracts were remapped as parts of the African Surface.

West Africa: From Senegal to Cameroon

The scheme of land surface development constructed by King was not applied directly to West Africa in work by francophone authors. There, the independent work of Vogt (1959), Michel (1973), Egbogah (1975), Grandin (1976), and Boulangé (1984) has left a lasting imprint on how landscape evolution in sub-Saharan Africa is analyzed. Correlations have, however, been attempted. The African Surface has been identified with the *surface bauxitique*, or more locally named *surface de Fantofa* by French authors, while the *relief intermédiaire* (hereafter termed "intermediate level") has been considered to correspond to King's post–African I surface. *Continental terminal* deposits are products of the erosion of Africa's swells as they have risen during the past 30 m.y. They have been laid down in basins among the swells and on the lower slopes of the swells themselves. The *continental terminal* was first defined by Kilian (1931) for Saharan continental sediments of Miocene to Pliocene age. Since the introduction of the terminology, its usage has fluctuated, and ferruginized Cenozoic sediments of marine origin in the coastal basins of West Africa have been included as part of this formation. In this paper the term is restricted to post-Eocene and pre-Quaternary continental sediments with well-defined upper and lower stratigraphic limits, as defined by Kogbe and Dubois (1980).

Erosional features on the swells that have been involved in colluvial transport or have been occupied by streams carrying *continental terminal* sediments are assigned to the intermediate level, which is fairly well age-bracketed by the sediments in the Iullemeden Basin of Niger, where Vogt (1959), Bocquier and Gavaud (1964), Gavaud (1966), and Boudouresque et al. (1982) all recognized that a topographic surface on weathered basement graded into a laterite-capped surface cutting *continental terminal* red beds. This topographic ramp between an erosional and depositional area on the edge of the Iullemeden Basin west of 5°E is interpreted as showing that the intermediate level is younger than the African Surface and that it developed as a consequence of the first manifestations of swell uplift and swell-flank erosion. The intermediate level is less well preserved on the eastern side of the Iullemeden Basin, where swell uplift in the Aïr was greater than that on the Guinea uplift to the west. Uplift of the Aïr generated a clear angular unconformity between the Eocene and post-Eocene strata (Faure, 1966; Boudouresque et al., 1982).

Colin et al. (2005) dated potassium-bearing manganese minerals in laterite from Tambao in Burkina Faso (15°N, 4°W) using $^{40}Ar/^{39}Ar$ laser-probe analysis of pisoliths and lateritic crusts. They concluded that greenhouse conditions, favorable to bauxite formation, characterized late Paleocene to early Eocene times (56–47 Ma) at Tambao and show a *bauxitic surface*, presumably the African Surface, flexed down over a distance of 750 km from an elevation of 700 m at Bondoukou close to the crest of the Guinea swell in Côte d'Ivoire (Grandin, 1976) (Figs. 14–17; see also Figs. 10 and 11) through a (now eroded) elevation of 500 m at Tambao to 200 m in the Niger Basin (Colin et al., 2005). They suggested that conditions on what they called a *ferruginous transitional lateritic land surface* (200 m lower than the bauxitic surface at Bondoukou) that outcrops at Tambao were dry at the beginning of Oligocene time (34 Ma), but reverted briefly to more humid conditions and the formation of Mn weathering products that yielded isotopic ages of 24–27 Ma. Colin et al. (2005) show their *bauxitic surface* as overlain by *continental terminal* deposits in the Niger Basin (Fig. 15). They show the *continental terminal* deposits as themselves overlain in the Niger Basin by their ferruginous transitional lateritic land surface, perhaps indicating a short interval (34–28 Ma?) involving local deposition of sediments of the *continental terminal*.

The widespread *haut-glacis* (hereafter termed "higher pediment") of West Africa was correlated with the post–African II (alias "widespread") surface of King (1976) and the *moyen-glacis* and *bas-glacis* (hereafter termed "middle" and "lower pediment," respectively) were regarded as probably Quaternary in age. The extent of the development of the middle and lower pediments in any one area was considered to have depended on local lithologic, paleoclimatic, and drainage conditions. Any flat-lying and bauxitized topography at elevations higher than the bauxitic surface was interpreted to be necessarily older than the Late Cretaceous, and therefore a post-Gondwana (sensu King, 1951) residual. That was considered to be the case for the Fouta Djallon, where authors (e.g., Michel, 1973) identified two high-elevation bauxitic levels: the Labé (≤1.4 km a.s.l.) and the Dongol Sigon (≤1.15 km a.s.l.) surfaces (Fig. 15). A more recent field survey of the Fouta Djallon in Guinea (Chardon et al., 2006) has suggested that the bauxite caps on both these surfaces, which differ in elevation by ~250 m, are probably distinct generations, the higher corresponding conventionally to King's post-Gondwana, and only the lower to the African Surface.

Because of the relatively inconclusive nature of their data, however, the authors are cautious in their preference for this

Figure 14. Examples of the African Surface and its geomorphic avatars on the Guinea swell, West Africa (after Grandin, 1976). (A) Location of key areas in Côte d'Ivoire studied by Grandin (1976), with details enlarged in panels B, C, D (Blafo-Gueto), and E (Bondoukou). Oblique hatching: metavolcanic and metasedimentary schist belts. Gray indicates granite-gneiss. (B) Detail of topographic laterite hierarchy in southwest Blafo-Gueto near Afotobo, in which a bauxitic African Surface remnant occupies the most elevated site. (C, D) Section through the Blafo (C) and Gueto (D) type areas showing dominant lithology and generations of pediments cutting into the African Surface. Note that the presence of all levels and laterite facies in any one area of the landscape is not systematic (compare with the synthetic figure for West Africa, Fig. 16). (E) Hierarchy of lateritic levels in the Bondoukou area of northeast Côte d'Ivoire. Note that the bauxite and upper laterites (intermediate level, higher pediment) developed primarily on the iron-rich metasedimentary and metavolcanic rocks, but that the pediments extend across to relatively iron-poor granitic or gneissic outcrops—because metal oxides were redistributed more widely across the landscape as it incised over time. This is true for most of West Africa. In B and E, most of the higher pediment is indurated in its upper- or mid-slope area, and the contact between the intermediate level and higher pediment is in many places marked by a scarp or "breakaway."

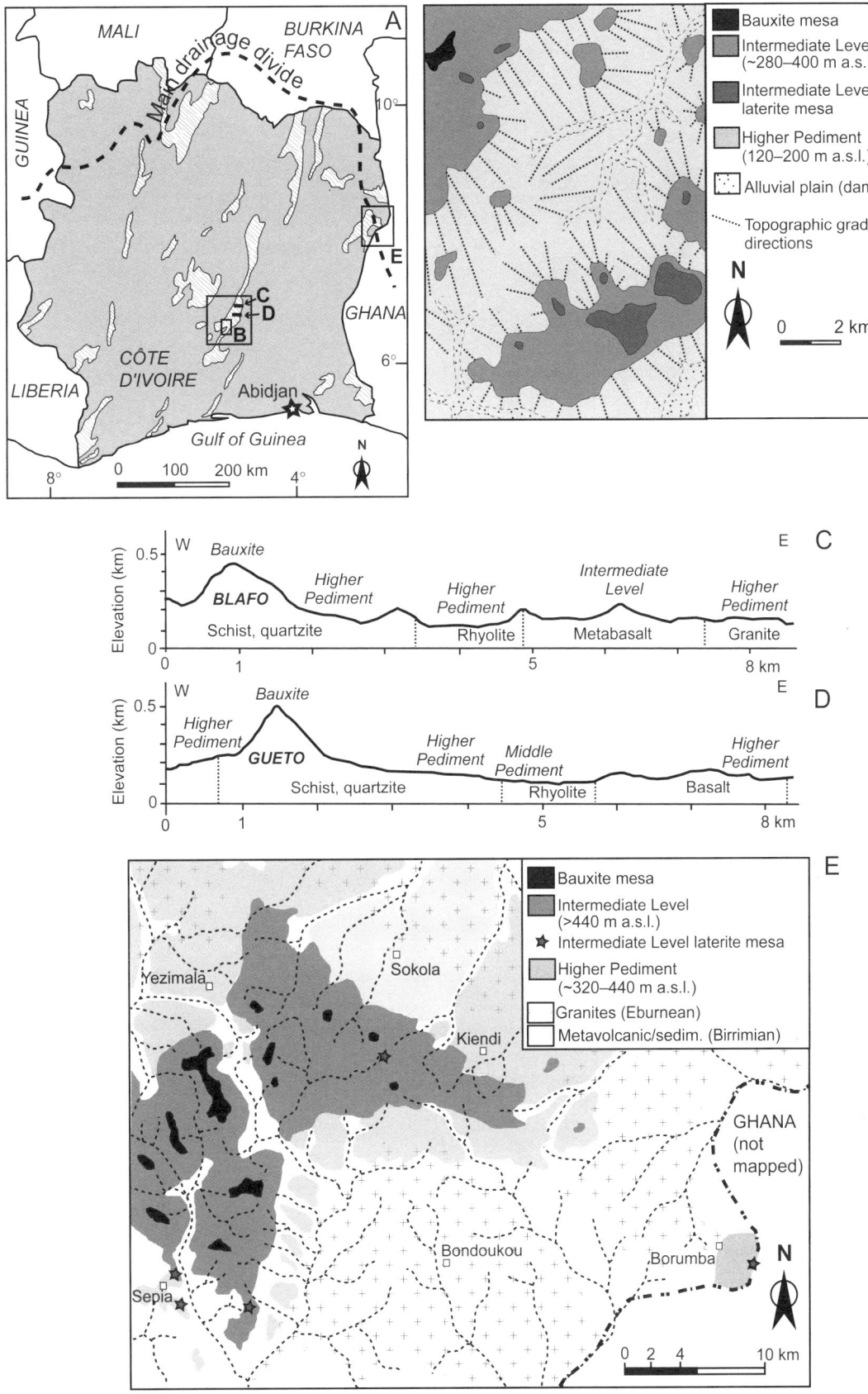

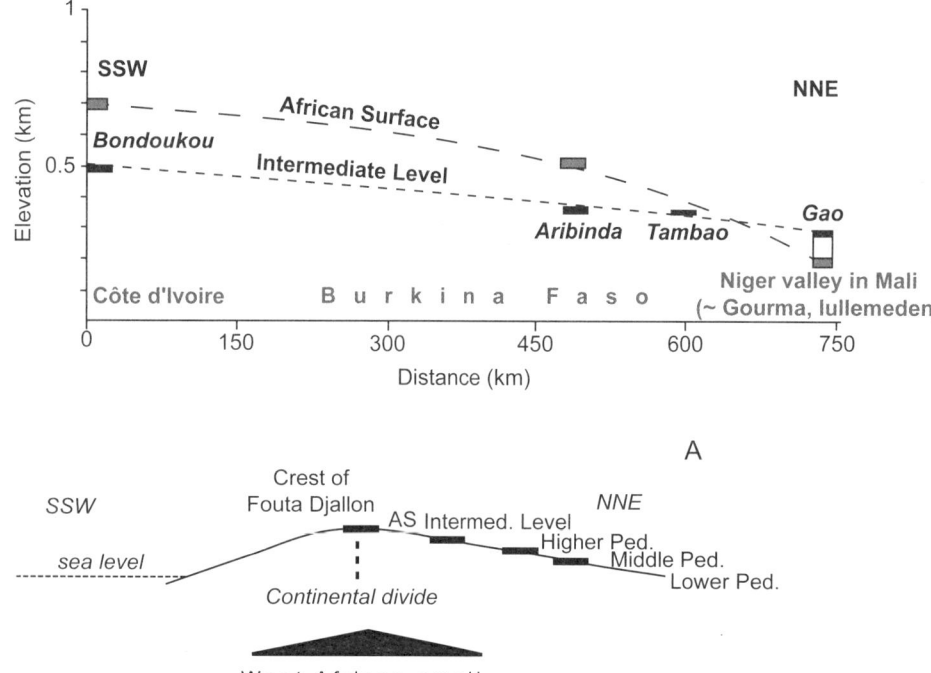

Figure 15. Post–30 Ma deformation of the African Surface on the northern flank of the Guinea swell based on dated manganiferous laterite outcrops (modified after Colin et al., 2005). Localities refer to correlated petrographic facies after Grandin (1976). Cross section located on Figure 2. See Figures 14A and 14E for location of Bondoukou.

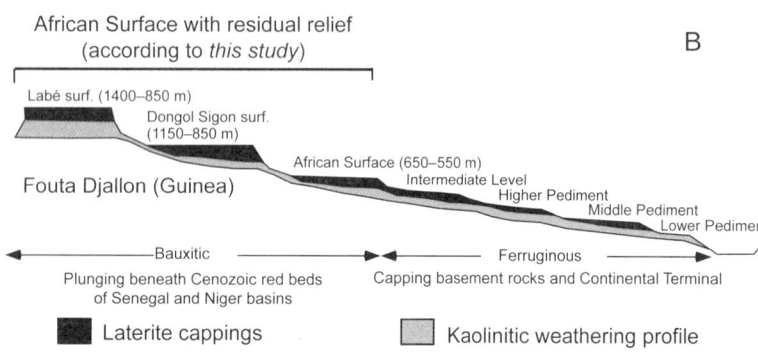

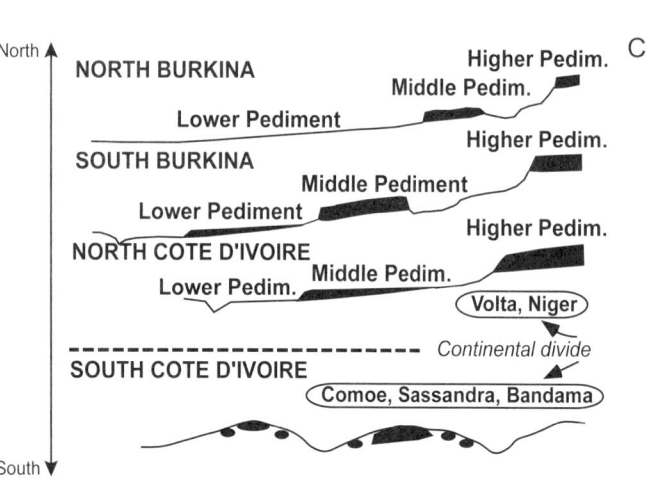

Figure 16. Schematic sketches showing the African Surface on West African swells and younger lateritized pediment surfaces that form stairways on the swell flanks. Cross section A located on Figure 2. (A) Asymmetric topography of the swell with Fouta Djallon crest zone corresponding to residual relief on resistant rocks standing proud of the uplifted African Surface. (B) Enlargement of (A) lateritic surface of West Africa (after Michel, 1973). Here we consider that the African Surface was uplifted after 30 Ma and that the seismically active Guinea highlands, which like the African Surface are bauxitic, formed residual relief and rose with the swell. There is no reason to consider that the Labé and Dongol Sigon bauxitic plateaus are the remains of erosion surfaces older than the African Surface. (C) Swell asymmetry has caused contrasting degrees of landscape rejuvenation by drainage incising into the swell flanks. As a result, the relative spatial extent of each recognized level (higher, middle, and lower pediments, for instance) and its laterite caprock varies according to position on the swell north and south of the continental divide. Here the succession corresponds to a north-south transect between the Gulf of Guinea (southern Côte d'Ivoire) and the Sahel (northern Burkina Faso). Partly after Tardy and Roquin (1998).

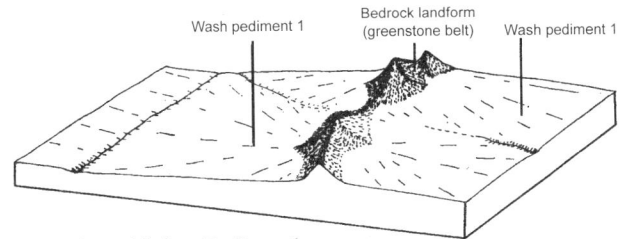

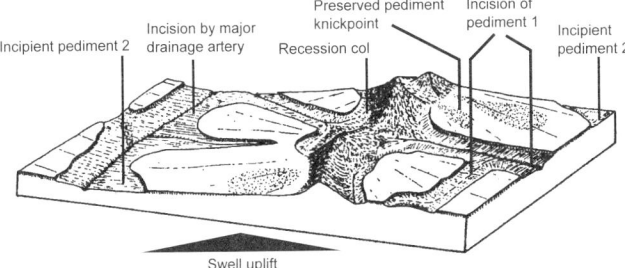

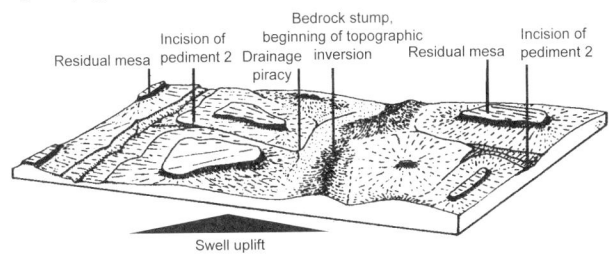

Figure 17. Example of laterite-capped pediment stairway formation in the granite-greenstone terrain in northern Côte d'Ivoire and Burkina Faso. Modified after Eschenbrenner and Grandin (1970).

interpretation, admitting that more work is needed. Furthermore, they did not consider the possibility, emphasized here, that the surfaces themselves are probably largely structurally controlled and relate to the low dips of the interlayered sandstones and quartzites. In keeping with the standpoint and supporting evidence presented in this paper (see earlier sections), we suggest that the two high surfaces do not correspond to separate cycles of denudation (such as a Gondwana and a post-Gondwana, or a post-Gondwana and an African cycle). The escarpment separating the two coeval surfaces appears instead to be controlled by structure and lithology, including thick doleritic sills (Bertrand and Villeneuve, 1989) probably related to the CAMP plume and injected into the metasediment pile ca. 201 Ma (e.g., Marzoli et al., 1999).

For those reasons, local rather than continental significance should be assigned to the Labé and Dongol Sigon surfaces. Evidence similar to that lately found in the Fouta Djallon, and indicating that an erosion surface has developed simultaneously at two different levels separated by a structurally or tectonically controlled step has been found elsewhere on the African Surface. In Burkina Faso, the escarpment separating the Taoudeni (latest Precambrian) sandstones from the older Precambrian basement (Gunnell, 2003, Fig. 1 therein) is an extremely ancient landform. Tiered lateritic mesas developed on the higher ground underlain by sandstone and on the basement rocks of the lowland duplicate the records of each of two erosion cycles operating simultaneously at two different levels—not a sequence of four cycles as was suggested by Daveau (1960). An analog to this situation is to be seen at the Arnhem Land sandstone escarpment in northern Australia, which seems to have been an upstanding landmark (a topographic island) at least since an early Cretaceous marine transgression left shoreline deposits at its foot (Needham, 1982; Frakes and Bolton, 1984).

The elevated Jos Plateau of Nigeria (Burke, 1996, Fig. 29 therein) constitutes a stepping stone between the high surfaces in Cameroon and the Guinea-Burkina bauxite-bearing surfaces 2000 km to the west. Bauxites belonging to the African Surface that occur on the Jos Plateau above an elevation of 1.25 km and are assumed to be of Paleogene age (65–22 Ma), have been described by Boulangé and Eschenbrenner (1971) and by Valeton and Beissner (1986).

A range of elevated, bauxite-capped outliers other than the Fouta Djallon and the Jos Plateau, such as the Nimba and Loma Mountains in central and eastern Guinea (Leclerc et al., 1949; Jaeger, 1953; Leclerc et al., 1955; Pascual, 1988), have also been used to correlate the African Surface across West Africa and to reconstruct patterns of post–30 Ma uplift (Egbogah, 1975; Grandin, 1976; Burke, 1996, Fig. 29 therein). The Jos bauxites can be tentatively correlated with those of the African Surface on the Guinea uplift and with the bauxitic Minim-Martap and Bamileke surfaces of Cameroon (Tardy and Roquin, 1998), which themselves occur at ~1.4 km a.s.l.

Another key indicator for West Africa is the reported presence of bauxite on the summit surface of the Hoggar (Patterson et al., 1986). The Hoggar is crowned by an extensive and elevated (2 km) erosional surface truncating both Precambrian metamorphic rocks and quartzites of the Precambrian basement, as well as unconformable outliers of Cretaceous nonmarine sandstone (Patterson et al., 1986). The sandstones and the basement are themselves locally covered by ca. 30 Ma and younger basalt flows (Dautria and Girod, 1991), indicating that the surface buried beneath the basalt in this area is the African Surface. An outlier of bauxite capping the Cretaceous sandstone has been reported to be buried by lava flows (Patterson et al., 1986). The flows descend toward the Tanezrouft Basin some 100 km to the southwest across unfaulted terrain, providing a measure of post-Eocene (34 Ma and younger) updoming (see also Burke et al., 2003a, p. 50). The elevated region has been affected by the eruption of younger trachytes and phonolites. At 2.7 km a.s.l., these now form the Hoggar swell's highest topographic points.

Burke (2001) reviewed the origin of the volcano-capped swells of the 1000-km-long Cameroon line. The individual Cameroon swells, half of which lie offshore, are an order of magnitude smaller in area than other individual African swells. Burke concluded from published isotopic ages of basaltic rocks that all ten of the ~100-km-wavelength volcano-capped swells began to rise ca. 30 Ma. The larger swells of the region that do not form parts of the Cameroon line, including the Ngaoundere and Biu swells, are more typical in length-scale of the general population of African swells. Dated volcanic rocks suggest that those swells also began to rise ca. 30 Ma. The short wavelengths and great elevations achieved on the separate swells of the Cameroon line have made working out what happened to the African Surface during the past 30 m.y. in Cameroon less clear-cut than has proved feasible on the gently upwarped swells of the West African craton. As in Kenya and Uganda, the correlation of erosion surfaces with those in adjacent regions that do not share a history of Cenozoic faulting and volcanism is difficult, and has led to diverse interpretations (Ségalen, 1967; Fritsch, 1978; Morin, 1989).

The Bamileke Plateau and Bamboutos hills in Cameroon are covered by latest Eocene (ca. 34 Ma) volcanic flows of the *série noire inférieure*, consisting of basalts, trachytes, tuffs, and ignimbrites. They seal a surface that is here correlated with the African Surface. According to Ségalen (1967), who, exceptionally among French authors, used L.C. King's surface classification most literally, the summit level of the Bamileke Plateau (1.45–1.2 km a.s.l.) and the summit level of the Adamaoua Plateau of north-central Cameroon constitute the "Minim-Martap surface." Above this level, which represents a partially bauxitized African Surface (Belinga, 1972), the more elevated plateaus in the far west of Cameroon are occupied by the Mambila surface, also known as the "Grassfields" surface, which extends into the Bamenda highlands of Nigeria and rises to 1.75–2 km. Altitudinal correlation suggests that the Minim-Martap and Mambila surfaces are of different ages, but Burke's (2001) analysis of the origin of the swells of the Cameroon line has indicated that the Cameroon line, because of the short wavelength of its individual swells, is probably one of the places in Africa least suited to the application of the altitudinal correlation method for distinguishing surfaces.

Morin (1989) also emphasized active tectonism on the Cameroon line. He identified a system of horsts in this part of Cameroon that he suggested played a part in elevating the volcano-capped swells by offsetting the African Surface since the beginning of Oligocene times (ca. 34 Ma). This is particularly relevant where the African Surface is overlain by younger Miocene (10–6 Ma) basalts. In contrast, the Meiganga-Ngaoundere (or Meiganga-Bamoun) surface, which is the most uniform and widespread surface of Cameroon, lies at an elevation of ~1 km, i.e., 200 m below the Minim-Martap surface. It, too, is overlain in places by unbauxitized Miocene basalts (Fritsch, 1978). At a lower level the *surface intérieure* (700 m), which is well defined on the basis of elevation, extends eastward into the Central African Republic as the *surface centrafricaine* at the same elevation (Boulvert, 1996), but is of undefined local age.

In summary, Ségalen (1967) followed an age-bracketing scheme based on current knowledge that assumed that the Bamileke surface was a Jurassic legacy, i.e., a relic of a Gondwana surface, following King. Fritsch (1978) was probably more correct in dating the Bamileke surface as Eocene, which would make it part of the African Surface. He placed many of the events that followed the beginning of upwarping of the African Surface within the Quaternary. That seems unlikely. Michel (1973), working in West Africa at about the same time, had also assigned many events to the Quaternary. Morin's (1989) recognition of horsts, which is consistent with Burke's (2001) analysis, is probably the most accurate work in Cameroon, although the chronology of block-faulting in relation to the successive volcanic eruptions and laterite-forming episodes remains to be unraveled. The only certainty appears to be that the African Surface in Cameroon predominantly occurs at an elevation of 1.45–1.2 km. It is probably considerably more upwarped on the individual volcano-capped swells.

Central Africa

In central Africa, the African Surface crowns the Nile-Congo and Chad-Congo divides at elevations little more than 0.5 km on the north side of the Congo Basin and the Congo-Zambezi divide on the south side at an elevation of 1.2–1.4 km. We recall that this is also the elevation of the Minim-Martap surface of Cameroon on the northwestern side of the basin. In the interior of the basin, the African Surface cuts across gently dipping Cenomanian and Turonian (100–90 Ma) terrigenous beds (Kwango series, comprising the Inzia sandstones and red argilites, and the soft Nsélé sandstones (Cahen, 1954, 1983).

Here the African Surface has been suggested to have evolved as a consequence of two major marine regressions in the South Atlantic, the younger having occurred in the Late Cretaceous (Lepersonne, 1960). The African Surface is unconformably overlain by outcrops of detrital sedimentary rocks that have been assigned to the poorly defined "Kalahari System sensu lato." These deposits, which represent episodic sandy and clayey outwash into topographic lows on the African Surface have been considered to have ages from early to late Tertiary and to be equivalent to the *continental terminal* of north-central and NW Africa. As in NW Africa, where both an earlier and a later Cenozoic *continental terminal* depositional pulse have been distinguished (Boudouresque et al., 1982), two sequences of terrigenous deposits attributed to semiarid conditions, both involving large lakes, have been identified. The lakes probably served as base levels for the development of local Cenozoic erosion surfaces. The first sequence is known as the *grès polymorphes*, labeled as Ba1 on geological maps because it has been widely observed across the Batéké Plateau of the Congo (Le Maréchal, 1966; de Ploey et al., 1968). These sandstones and clays have been correlated with similar formations in South Africa on the basis of fossiliferous marine intercalations, but the age remains moot (Cahen and Lepersonne, 1952). The second sequence, which is stratigraphically younger than the *grès polymorphes* and is also widespread on the Batéké Plateau, is known as the *série des grès ocres*, or Ba2. These represent a reddened facies of kaolinitic sandstone that marks a return to more humid conditions in central Africa, perhaps related to the establishment of the

South Atlantic permanent anticyclone and its moisture-advecting monsoon systems after Eocene times. Although the ages of the Cenozoic nonmarine sedimentary rocks of the Congo Basin are poorly defined, their occurrence on the lower slopes of the African Surface, where the surface dips downslope into the lowest parts of the basin, indicates that they, like the *continental terminal* deposits, are likely to have developed after the upwarping of Africa's swells began ca. 30 Ma (Petit, 1994). The occurrence of extensive lakes occupying large parts of the Congo Basin provides another analog with the Chad and Kalahari Basins, which have both been partly occupied by extensive lakes at times during the past 30 m.y.

A 1000-km-long N–S-trending range, the Mayombe mountain range, separates the Congo Basin from the Atlantic Ocean, forming an actively rising swell on the western flank of the basin. Seaward of those mountains, laterite-capped terrigenous red beds, equivalent to the *série des grès ocres* and 30–180 m thick, are known as the *série des cirques* (Giresse et al., 1981). They are possibly linked to a marine regression during the latest Miocene, but perhaps represent a longer period of deposition. The beginning of swell uplift ca. 30 Ma shows itself on the Atlantic seaboard flank of the Mayombe Mountains in an intra-Oligocene (34–22 Ma) lacuna in sedimentary sequences from Gabon to northern Angola. That uplift was accompanied by a well-documented seaward flexure (see, e.g., Marton et al., 2000) in which the eastern edge of the marine basin emerged above sea level. Cutting of deep valleys and submarine canyons (Marton et al., 2000, Pl. 4 therein) accompanied the Oligocene regression. This fall in sea level is attributable to both the growth of the Antarctic ice sheet, starting ca. 34 Ma, and the beginning of the development of Africa's swells at 30 Ma.

High volumes of deep-water sediments of Oligocene (34–22 Ma) and younger age in the offshore basins of Equatorial Guinea, Gabon, Congo, and Angola confirm the timing of initiation of the rise of the Mayombe Mountains. The most spectacular deepwater sediment body in the region is the Congo Fan, which extends over a ~2 × 10^6 km^2 area of the Atlantic Ocean floor. The Congo fan began to develop in Oligocene time (ca. 30 Ma), with documented sediment accumulation rates increasing through the Neogene (Mougamba, 1999; Séranne, 1999; Lavier et al., 2001; Anka and Séranne, 2004). Cenozoic denudation in the Congo and the smaller coastal river catchments accounts for the bulk of terrigenous sediments accumulated during that period (Leturmy et al., 2003). The total volume of that sequence was estimated at 1.2 × 10^6 km^3, making up half of the total post-rift sequence (Leturmy et al., 2003). The thickest (>4 km) and largest (>0.5 × 10^6 km^3) post-Oligocene depocenter lies across the margins of Congo and Angola, and extends over the oceanic crust to the Congo deep-sea fan. At a maximum, the fan has accumulated an estimated thickness of ~3 km of sediment (e.g., Tari et al., 2003, Fig. 14 therein). Earlier Congo fan deposits were fed by a river, or rivers, that reached the coast in an area 200 km north of the present mouth of the Congo where the Kouilou/Niari river system reaches the coast today (Uenzelmann-Neben, 1998).

Because the volume of sediment in the Congo fan is so large, it seems likely that the Mayombe Mountains were breached by through-going rivers relatively early during the past 30 m.y. This can be reconciled with the existence of large freshwater lakes in the Congo Basin during the same interval and the idea that these lakes spilled into rivers that had cut through the Mayombe Mountains. One possibility, by analogy with the history of the Chad Basin over the same interval, is that lakes in the Congo Basin were at times isolated and at other times part of drainage systems that reached the Atlantic. Because the Intertropical Convergence Zone has played such a dominant role in the meteorologic history of the Congo Basin since the East Antarctic ice sheet first formed ca. 34 Ma, rainfall has probably always been high, and inland drainage may not have been important.

In a model similar to those of Driscoll and Karner (1994) and Whiting et al. (1994) for the Amazon and Indus fans, respectively, Lucazeau et al. (2003) speculated that by applying the Congo deep-sea fan load at a distal position with respect to the margin, the result would increase the subsidence of the offshore margin and generate 100 m of maximum flexural uplift along a strip parallel to the coast, several hundreds of kilometers inland. Such flexural uplift would have favored erosion of the coastal catchments and contributed sediment to the margin. However, such low magnitudes of uplift are probably insufficient to account for the Neogene sediment fluxes, which is why we advocate that dynamic swell uplift after ca. 30 Ma is the key mechanism responsible for the geodynamic evolution of the Mayombe Mountains.

Uenzelmann-Neben (1998) concluded that the change from more northerly drainage through the Mayombe Mountains to the present Congo River occurred during late Pliocene times. Petit (1994) had suggested that the course of the present Congo was caused by drainage of a small tributary to the Kimpoko Pool being reversed by a small Atlantic river cutting back through the Mayombe Mountains. That event was postulated to have coincided with a marine regression in the Atlantic. This lowered base level triggered headward ingress of drainage into the continent, possibly involving one of the last outwash pulses of a late Miocene to Pleistocene terrigenous *série des cirques* (Massengo, 1970) in the coastal regions of Gabon and Angola. Headward valley cutting had, however, already occurred several times since the early Oligocene in response to uplift of the Mayombe Mountains and related uptilting of strata on the continental margin.

This evidence gives little credence to Petit's (1994) claim, after Cahen (1954), Lepersonne (1960), and apparently Veatch (1935) before them, that the late Pliocene or Quaternary drainage diversion across the Mayombe Mountains due to headward capture by an Atlantic stream was the first of its kind. It also casts doubt on the belief that, before this drainage diversion toward the Atlantic, the Congo Basin rivers connected to the Chad Basin over the low Central African Republic swells (e.g., Cahen, 1954). A vast literature, much of it petroleum related and not necessarily all in the public domain, on the oil-producing Congo fan off Angola reveals that the unconformity at the base of the fan is

30 Ma in age. The Congo fan has been fed by a river or rivers draining the Congo Basin through at least two different river channels through the Mayombe Mountains since 30 Ma. The late Pliocene (ca. 2 Ma) episode appears to have been responsible for the initiation of the present Congo submarine canyon, which is 230 km long (44 km is cut into the continental shelf), and reaches depths of 2.3 km following a slope of ~1‰ (Babonneau et al., 2002). The current drainage system of the Congo was established across the downwarped complex of Cretaceous to Holocene deposits on a multifaceted African Surface within the Congo Basin. That surface is occupied by dune fields, outwash plains, lake floors, and, on the upwarped edges of the basin, stripped vestiges of the African Surface that carry no cover.

On the northern rim of the Congo Basin, the Central African Republic is dominated by the *surface centrafricaine* (Boulvert, 1996). The territory corresponding to the Central African Republic represents a topographic saddle between the East African rift shoulders and the Cameroon swells. The *surface centrafricaine* is found within an elevation bracket of 0.55–0.7 km a.s.l. across a 2000 km, E–W-trending tract and is upwarped to a maximum of 0.9 km on the west and the east as it approaches the Cameroon and East African swells. As such, on the basis of altitudinal criteria, it has been correlated to the west with the *surface intérieure* (0.7 km) of Cameroon and to the south with the Batéké Plateau of Congo. The *surface centrafricaine* slopes gently northward toward the Chad Basin (at 0.3–0.4 km a.s.l.), and southward toward the Congo Basin, forming two asymmetric piedmont surfaces on which sedimentary sequences have been accumulating intermittently between Cretaceous and Pleistocene times.

The *surface centrafricaine* in the Central African Republic thus links higher ground to the east and to the west, but it also forms a low continental drainage divide between the Congo, Chad, and Nile basins. The surface carries two higher, less extensive surfaces and a sandstone unit that help in reconstructing the denudation chronology of Central Africa. In the western part of the Central African Republic, the elevated Lim-Bocaranga and Bouar-Baboua surfaces at ~1.2 km and 1 km a.s.l. have been interpreted, purely on altitudinal grounds, as outliers of the Minim-Martap and Meiganga-Ngaoundere surfaces of Cameroon (Boulvert, 1996). Other elevated areas of the Central African Republic correspond to two large sandstone plateaus crowning the *surface centrafricaine*. They correspond to the sporadically distributed Gadzi and Wadda Cretaceous sandstone formations and represent fluvial material routed northwestward from an eroding highland to the south. They correlate stratigraphically with the Cretaceous *continental intercalaire* of the Sahara and Sahel regions. The sandstones are capped with laterites. There are lateritic levels on both the sandstone and less elevated underlying surface on basement rocks. This relationship is similar to those observed in the Fouta Djallon and western Burkina Faso (cf. *supra*).

Clues toward establishing a chronology of surfaces in the Central African Republic have remained elusive. As a result, correlation of the laterite capping the Congo-Chad Basin divide with the West African models of Michel (1973) and Grandin (1976) remains speculative. Altitudinal correlations of surfaces in the Central African Republic with those in Cameroon and Uganda have proved difficult even with the help of laterite fingerprinting. Correlating regolith characteristics remains particularly difficult because of the presence of bauxite in Cameroon and its absence from the Central African Republic and Uganda, so that bauxite cannot be used with confidence as an index marker-rock. It is well established that unweathered basalt generates bauxite (a silica-ridden weathering product) faster than do unweathered basement rocks (Grandin and Thiry, 1983) because there is no quartz in basalt. For this reason, the stepped landscape, well characterized in its West African type area is, in the Central African Republic, collapsed into a single land surface poorly defined in terms of age. Here we suggest that the widely bauxitized Minim-Martap (Cameroon) and Buganda (Uganda) surfaces are parts of the composite African Surface. Differences in elevation that exist among those surfaces are attributable to faulting, warping, and lithological controls, differences in bauxite occurrence being due to paleoclimate (see later section on paleoclimate). Boulvert (1996) emphasized that no clear-cut staircase of surfaces is observed in the Central African Republic, and preferred to speak of a composite post-Cretaceous topography in which surfaces grade into one another. Only in the northwest, near the Cameroon border, has faulting accentuated topographic contrasts. The *surface centrafricaine* in the Central African Republic meets our definition of the African Surface as a composite surface.

Farther to the south, on the northern edge of the Congo Basin, the surface corresponding to the erosional cycle subsequent to the African surface (the "Victoria Falls" cycle of King, alias his post–African I surface) is buried by the late Cenozoic fluvial and lacustrine *sables ocres* (Ba2), which are frequently lateritized. These are contemporary with the *continental terminal* of West Africa (latest Eocene to Oligocene, post–34 Ma to 23 Ma). The lacustrine nature of these outcrops testifies to the senility of the topography that they underlie and to the very shallow depths of postdepositional denudation. Because of the limited amount of subsidence of the Congo Basin during the Cenozoic, the African Surface has been considered to merge with an older "post-Gondwana" surface that is sealed by the Upper Cretaceous Kwango beds. This older surface dips at a low angle (1.6% slope) toward the NW in Kwango-Kasai on the south side of the Congo Basin, where it is exhumed from beneath its Upper Cretaceous cover. Whether in Africa or around the Hercynian swells of western Europe (Klein, 1990; Godard et al., 2001), it is typically in transition zones in which topographic swells grade into sediment-filled basins that successive generations of topographic paleosurfaces are most difficult to distinguish from one another. They grade into one another at low angles so that elevation differences and even contrasts in weathering mantles may not always be sufficient to constrain paleotopographic reconstructions. Such piedmont zones constitute composite surface territory *par excellence*. The difficulty in clearly distinguishing between the African Surface and other features of African landscape development such as a "post-Gondwana" surface of a different age perhaps justifies the position taken in this study, which treats the African Surface as the end product of composite cycles of erosion that started in the Middle Jurassic (ca. 180 Ma) or Early Cretaceous (ca. 125 Ma) with continental breakup events and lasted for up to 150 m.y.

In southern Zaïre (Shaba Province of current Democratic Republic of Congo), the Manika Flats and the tops of the Kundelungu Mountains (1.8 km) are considered to represent remnants of the African Surface, which bevel both the crystalline basement and the folded Neoproterozoic Katanga Group metasedimentary rocks. Some of these mountains appear to occupy horsts flanking the Lake Upemba graben (at 9°S, 26°E). Those horsts are capped by silicified lacustrine limestones of unknown age but perhaps correlatable with the lacustrine *grès polymorphes* (post–30 Ma?) of the Congo Basin. The evidence of preserved lacustrine sediments testifies to limited depths of denudation since uplift across these elevated land surfaces. Different generations of late Cenozoic surfaces have been identified on topography below the Manika and Kundelungu Plateaus on the basis of Kalahari-type cover sands and the contrasting nature of iron oxide coatings on laterites (Alexandre and Alexandre-Pyre, 1987). Farther south, in northern Zambia, the African Surface declines in elevation to ~1.15 km a.s.l. in the Lake Bangweulu depression.

In the northern Democratic Republic of Congo, lateritic surfaces have been correlated (Lepersonne, 1956; King, 1962) with those of the closest well-studied area in Uganda, but no recent progress has been reported either in dating the laterites or in correlating the petrographic facies of the ironstone caprock (Dumont, 1991). North of the Nile-Congo divide, in southern Sudan, Schwarz (1994) mapped several laterite-capped mesas developed on Cretaceous sedimentary rocks in Jebel Howag and in the Nuba Mountains near Kau that we suggest represent parts of the African Surface.

East Africa: The Great Swell and the Rift System

Intensely faulted rift valley relief, although a major focus of geological investigation for more than 100 yr, occupies less than one-quarter of the surface area of East Africa. The remainder of the region, including eastern Kenya, central and northern Uganda, and western, central, and southern Tanzania, consists of a single great swell: the East African swell, exposing extensive tracts of deeply eroded Precambrian rocks. Uganda has provided a casebook for geomorphic surface reconstruction in East Africa from the time of Wayland (1933, 1934) and later Bishop (1966; Bishop and Trendall, 1967). Subsequent analyses have produced a picture almost as confusing to outsiders as is that of the South African type area of the African Surface (see above).

Three tiered surfaces have been mapped and tentatively age-bracketed according to the most up-to-date volcanic and paleoclimatic data. Taylor and Howard (1998) contend that the lateritic mesa surfaces that cap the highest summits of Uganda at 1.45 km a.s.l., suggested to be of Jurassic to mid-Cretaceous age, are supported by resistant metasedimentary lithologies. The surface corresponding to these summits was called the Koki surface by McConnell (1955). It was called the post-Gondwana surface by King (1962) and the Cretaceous surface by Dixey (1956). The surrounding, stripped surface at 1.2–1.3 km a.s.l. was eventually suggested by King (1962) to be the African Surface, and it is still usually identified with that surface (Doornkamp, 1968, 1972). The stripped surface, which lies <250 m lower than the Koki surface, is also frequently referred to as the Buganda surface because it is commonly preserved on quartzites of the Buganda series. The higher Koki surface is developed on Karagwe-Ankolean metasediments, so that the escarpment separating the Koki from the Buganda surface has been regarded by some as not of cyclic origin, i.e., caused by base-level change, but as simply related to the contrasting susceptibilities to weathering of the two bedrock types (Pallister, 1960; Bishop, 1966). We consider that this elevation difference is of local rather than continental significance, similar to that discussed earlier for the Fouta Djallon in West Africa.

The Buganda surface may therefore be considered to be the African Surface of this paper: it carries residual topography (of 250 m) that has survived only as a result of lithological contrast. Ollier (1960, 1981: p. 159) and Ollier et al. (1969) did not distinguish the Koki surface and considered that the Buganda surface was the most elevated mappable surface in Uganda, and for that reason an equivalent of the Gondwana surface of King (1951). Ollier even more confusingly labeled as "Africa surface" [*sic*] the areally extensive "Tanganyika" or "Victoria Falls" (King, 1962) or "Kyoga" (McFarlane, 1976) surface of Dixey (1956) and Pallister (1960)—which is more commonly identified with the post–African I surface of King. The African Surface, in this area called the Buganda surface, was locally stripped following drainage expansion from base levels to the west within the Western Rift system during Miocene times. Burke (1996, p. 383) concluded that presently available data indicated an origin for the Western Rift system ca. 15 Ma. That yields a time of 15 Ma for the beginning of the stripping of the African Surface by westward flowing streams.

A supporting argument for attributing an Eocene to Oligocene terminal age (53–24 Ma) to the African Surface is that it is covered by (1) Miocene volcanic flows (Bishop, 1958) and fossiliferous lake deposits in the Kavirondo Gulf of Lake Victoria, and (2) the volcanic rocks of Mount Elgon (erupted ca. 20 Ma) and in the Lake Turkana region near the eastern Ugandan border (Morley et al., 1992). Lateritization of the African Surface had occurred before the downfaulting of the Albertine graben ca. 13 Ma, indicating favorable climatic conditions at some time prior to that event. Apatite fission-track and helium data also indicate an absence of rock cooling between the main peak of denudation that occurred in response to the Santonian tectonic event in Late Cretaceous times and rift-flank uplift in the late Neogene (Spiegel et al., 2007). This strongly suggests an absence of local relief in the region until late Cenozoic times, hinting at the extensive presence of a low-relief African Surface across the region. The African Surface was uplifted again and stripped further during the mid-Pleistocene on the Western Rift flanks (Taylor and Howard, 1998).

In Tanzania, where few remnants of older surfaces have been identified, the African Surface is represented either by broad, even, interfluves or by wide plateaus such as the surface cut in gneiss across the Serengeti Plains at an elevation of 1.5 km a.s.l. Correlation with the African Surface was confirmed in some detail in neighboring Rwanda-Burundi, where Rossi (1980) surveyed four stepped surfaces notched into the eastern flank uplift of the Western Rift. He named the "Byumba" surface as an extension of the Jurassic to mid-Cretaceous (Koki?) surface reported from Uganda, and the "Butare" surface (1.7–1.8 km) as an extension of the African Surface. The latter is ubiquitously sealed by alumino-ferruginous laterite and dissected into flat-topped sinuous ridges.

The escarpment separating the Butare surface of Rwanda from the younger "Kagera" surface (1.4–1.6 km) is abrupt and deeply dissected as a direct consequence of rejuvenation by uplift of the rift shoulder to the west. Following anglophone writers in the neighboring regions, Rossi (1980) correlated the Kagera surface with the post–Africa I ("Victoria Falls") cycle of King, i.e., the Tanganyika surface of McConnell (1955) and the late Tertiary surface of Dixey (1956). Because (1) its continuation into Uganda is buried under the Pliocene Kaiso series of Lakes Edward and Albert, and (2) river terraces bearing late Pliocene to Quaternary hominid artifacts occur at slightly lower elevations, it seems safe to infer that the Kagera surface is of late Miocene age (>5.5 Ma). In summary, if the correlations by Rossi of the tiered surfaces of Rwanda-Burundi with King's surfaces in Uganda are correct, the greater elevations of the African Surface in Rwanda-Burundi (1.8 km) compared to Uganda (1.3 km) indicate greater magnitudes of Miocene and continuing rift-flank upwarp in the Lake Kivu and Lake Tanganyika areas to the south in comparison with flank uplift in the Lake Albert region.

Şengör (2001, p. 201–202 and Fig. 12 therein), in an approach that was new among students of the erosion surfaces of the East African swell, emphasized the likely influence on late Mesozoic and early Cenozoic relief of the Sudan and Anza rift shoulders (Wysick et al., 1990; Winn et al., 1993). Those rifts were initiated during the Late Jurassic (ca. 150 Ma) and were active, in the sense that sedimentary rocks were being deposited in them, episodically until the late Eocene (ca. 40 Ma). Şengör (2001) concluded that the Kenyan "end-Cretaceous" surface of Saggerson and Baker (1965), which we consider was part of the African Surface, rose gently from the east coast of Africa toward the Anza rift and that the Buganda surface of Uganda (Bishop and Trendall, 1967) and the early Cenozoic Kasubi surface of the same authors sloped gently westward away from the Anza rift and toward the Atlantic Ocean.

In the Congo Basin, that westward sloping surface has been dated by Cahen (1983) who found the Nsele Group of early to middle Cenomanian age (99–98 Ma) to be truncated by the surface on which the *grès polymorphes* have been deposited. Those sandstones, which are possibly of the same age as the *continental terminal*, have been dated on nonmarine faunas and floras as being of early Cenozoic (Haughton, 1963, p. 318) and Eocene–Oligocene age (Furon, 1960, p. 306–307). Şengör considered that the regional culmination in elevation corresponding to the NE–SW-trending Sudan and Anza rift systems would have been no greater than 0.5 km a.s.l., although he recognized that rift shoulders might locally have been higher. In this paper, we consider the occurrence of relief of up to 0.5 km above the general level of the African Surface pediments to have been likely at rift shoulders on the mature African Surface.

It seems generally that plateaus in East Africa at elevations similar to that of the Serengeti Plain (1.5–1.9 km) are remnants of the African Surface. Those plateaus include the eastern flank of the Lake Tanganyika Rift in western Tanzania, the Masai highlands of Kenya, the Muchinga Plateau of central Zambia, the Blue Mountains of the northeast Democratic Republic of the Congo, and the Manica and Harare highlands of Zimbabwe. Among the highly elevated summits of East Africa, most are: (1) young, normally faulted blocks that have rotated within the past 5 m.y. such as Ruwenzori; or (2) Neogene volcanoes such as Mount Kilimanjaro, Mount Kenya and Mount Elgon. Anomalously elevated massifs on the East African swell, with as yet uncertain links to the rift system, include the Uluguru Mountains of eastern Tanzania, where a highland surface on resistant charnockitic gneisses occurs at an elevation of 2.4 km in an area close to active faults of the eastern rift system. As in South India and Sri Lanka (Gunnell and Louchet, 2000), the great resistance to erosion of the charnockite may help to explain the anomalously high topography of the Uluguru Mountains over the long term.

The region containing the Rukwa Rift and the northern part of the Malawi Rift occupies the southern end of the East African swell (Burke et al., 2003b, Fig. 20 therein), where it is cut by rifts of the western arm of the East African rift system. The region has been recognized to expose rifts of three generations: (1) Rifts initiated between ca. 310 and 250 Ma during the final assembly of Pangea. Those rifts were initiated during the world's most widespread episode of intracontinental deformation of the past 560 m.y. (Burke et al., 2003b, p. 30). The rifts contain Permian-aged sedimentary rocks of the Karroo system best known in the Rukwa area (Morley et al., 1992). (2) The second generation are rifts that originated during the Karroo plume plate-pinning episode (Karroo-PIPPE, 183–133 Ma; Burke et al., 2003a and Fig. 14 therein), which in Malawi were most strongly expressed in igneous activity. The Shire rift was characterized by the eruption of abundant nepheline syenites and carbonatites with 140 Ma and younger ages (Burke et al., 2003a, Fig. 1 therein). Rivers flowing along the Shire rift and its downstream extension, the Urema rift, fed the Zambezi delta at the Mozambique coast throughout the Cretaceous. After a break from ca. 65 Ma to ca. 30 Ma, successor rivers again conveyed sediment to the coast of Mozambique (De Buyl and Flores, 1986; Burke, 1996, Fig. 47 therein). (3) The third-generation rifts were rifts of the active western arm of the East African rift system. The initiation of these is poorly dated but is thought to have occurred ca. 15 Ma (Burke, 1996, p. 385).

AFT studies (Van der Beek et al., 1998) have confirmed the three-stage tectonic history of this area and helped to clarify the denudation chronology by recording intervals of tectonism and erosion associated with all three rifting episodes. The summits of the Kipengere (or Livingstone) Mountains and the Lupa Plateau yield ages associated with final Pangean assembly recording cooling, within the apatite partial annealing zone (PAZ) but without surface exposure, between 250 and 200 Ma (Van der Beek et al., 1998, p. 374). Those results relate to a time of tectonism and erosion when Pangea was a single continent long before the African Surface-related erosion began. A population of ages indicating passage through the PAZ at 150 ± 20 Ma from the western flanks of the Malawi and Rukwa Rifts indicates tectonic activity during the Karroo-PIPPE while the Shire rift was active (Van der Beek et al., 1998, p. 374). Samples from the escarpment of the Kipengere range become progressively younger with depth and do not show PAZ influence until 60–70 Ma (Van der Beek et al., 1998, p. 374 and Fig. 5 therein).

Delvaux and Wopfner (1992) identified a continuous erosion surface on the summit of the Kipengere Range that we consider to be part of the African Surface. Van der Beek et al. (1998) regarded the concordant AFT ages and the fission-track length-distribution from that plateau to be consistent with its being capped by a continuous erosion surface. Comparable AFT length distribution signatures from the summit of the Lupa Plateau led the authors to suggest that the same surface is exposed in both areas although the Lupa Plateau is lower by ~1 km. Those same authors (1998, p. 374) considered on the basis of AFT that the surface capping the Kipengere Mountains had a maximum exposure age of ca. 40 Ma. That area is at the southern end of the East African swell (Burke et al., 2003a, Fig. 20 therein), which began to be uplifted ca. 30 Ma (Burke, 1996, Fig. 32 therein). Modeling of AFT track lengths (Van der Beek et al., 1998, Fig. 8 therein) for the Kipengere escarpment was consistent with most of the Cenozoic cooling in the area having taken place during the past 20 m.y. (Van der Beek et al., 1998, p. 378), thus matching the ~15 m.y. timing suggested by other evidence for the initiation of the western rifts (Burke, 1996).

In summary, the AFT results are consistent with long-term erosion in the Lake Rukwa and Lake Malawi regions between ca. 180 and ca. 40 Ma. Perturbation by the Shire rifting episode (150 ± 20 Ma) during Karroo-PIPPE times is recognizable on the western flanks of the Rukwa Rift and west of northern Lake Malawi, but not in the Kipengere Mountains. That rifting episode is known from the distribution of associated alkaline intrusions to have been better developed in the southern Lake Malawi area than near the Kipengere Mountains. The African Surface has been elevated to 2.6 km on the active rift shoulder in the Kipengere Mountains. Regionally it lies at ~1.6 km on the Lupa Plateau and over much of Mbeya Province in Tanzania. Farther south, on the active Lake Malawi rift shoulder at the Malawi-Mozambique border, the Mulanje Mountains expose what are here suggested to be exposures of the African Surface at 2 km a.s.l. Those exposures are capped by lateritic bauxite over syenites intruded during the Shire rifting episode. Those rocks have yielded K/Ar ages on biotite of 116 ± 6 and 128 ± 6 Ma (Garson and Walshaw, 1969). The ages show that the part of the African Surface on which the bauxite formed is younger than Early Cretaceous in age.

The spectacular elevation of Mulanje is a result of shoulder uplift of the currently active Malawi Rift, and therefore a local topographic anomaly of a kind common in the youthful parts of the East African rift system. That is compatible with the more general conclusion that the African Surface occurs at very high elevations only where it has been recently uplifted by young (post–15 Ma) rift-flank movements, and at lower, but still relatively high, elevations (typically 1–1.5 km) where it caps unrifted, post–30 Ma swells.

Northeastern Africa and Arabia

Occurrences of the African Surface identifiable by lateritic, and in a few places bauxitic, weathering are distributed sporadically over much of what had been, during the formation of the African Surface, the vast low lying area of northeastern Africa and Arabia. Minimum African Surface ages are in many places well dated as older than 31–28 Ma, which is the age of eruption of the Ethiopian traps. More precise ages exist in only a few cases. (1) Collenette and Grainger (1994) reported Late Cretaceous bauxites from northern Saudi Arabia. (2) Lateritic bauxites capping thin Coniacian (ca. 87 Ma) sedimentary rocks are reported from the Marbat area of southern Dhofar, suggesting emergence and low elevation of the southern Arabian peninsula in Late Cretaceous times (Roger et al., 1989). (3) Davidson and Rex (1980) and Davidson (1983) reported the presence of a basal unit of red lateritic grit below the 45 Ma basalts of the Omo River valley in southern Ethiopia, as well as beneath other younger basalts (ca. 30 Ma) elsewhere in that region. The laterite was interpreted as capping a low-relief pre-rift surface onto which the basalts were emplaced. (4) In the Wadi Natash region of Egypt (24°N, 34°E), a gibbsitic latosol is developed on Upper Cretaceous volcanic rocks (Saïd et al., 1976). (5) South of Sana, in Yemen, and immediately below the trap basalt (aged ca. 30 Ma), the top of the Medj-zir Formation of the Cretaceous to early Cenozoic marine to terrestrial Tawilah Group is occupied by ~100 m of laterite (Geukens, 1966; Al-Subbary et al., 1998). (6) In Eritrea, laterites that underlie the 30 Ma basalt "*represent a widespread and lengthy period of humid conditions in the [early Cenozoic] that developed on a near sea-level plain*" (Drury et al., 1994, p. 1372). The laterites cap basement rocks as well as the Jurassic Adigrat sandstones. (7) More generally, Bohannon (1986), Garfunkel (1988), and Briem (1989) reported the occurrence of a lateritized African Surface on both shores of the Red Sea, particularly emphasizing lateritic soils in Sudan and southern Saudi Arabia capping Proterozoic rocks. The soils are sealed by mid-Miocene (ca. 14 Ma) basalts. (8) The red beds of Suez and Sinai have been inferred to represent material eroded from the laterite-bearing African Surface during the Oligocene.

By plotting the distribution of marine shorelines, Şengör (2001, Fig. 12 therein) showed that the entire region of northeastern Africa and Arabia was low lying between the beginning of the Paleocene (65 Ma) and late Eocene times (36 Ma). The complex shapes of the shorelines and their rapid horizontal fluctuations, which in some areas exceeded 1000 km between early and late Eocene times (an interval of <20 m.y.) led Şengör to conclude that he was dealing with a very low lying area. Early Cenozoic (65–36 Ma) transgressions and regressions were on a low lying land surface that we identify as the African Surface. Merla and Minucci (1938, p. 347) had earlier described Ethiopia as having been an area of extremely low relief occupied by a surface of low elevation that was a product of "*erosione prevulcanica*" (i.e., pre-volcanic erosion) during Late Cretaceous, Paleocene, and perhaps also Eocene times. The volcanic activity to which Merla and Minucci referred was the eruption of the Ethiopian traps, which is now well dated as beginning at 31 Ma (Hofmann et al., 1997). In the light of the latest information, Merla and Minucci's surface can therefore be considered to have been low lying from ca. 95 Ma until close to 30 Ma. It corresponds to the surface that we call here the African Surface. Şengör (2001, Fig. 12 therein) also showed "*erosion, soil formation and local continental deposition*" as characterizing an area of ~10^6 km^2 in Sudan. We interpret this area to have been occupied by the African Surface during Paleocene to early Oligocene times (65–30 Ma).

The elevation of the Afar Dome, the associated eruption of the Ethiopian traps, the establishment of the Ethiopian Rift and the development of the Red Sea and the Gulf of Aden, first as intracontinental rifts and more recently as places in which ocean floor is forming, have led to complex and episodic topographic deformation of the African Surface. In many areas, the lateritic cover has been completely removed by erosion. Burke (1996, p. 372–387) provides the only modern comprehensive review of the evolution of the whole East African rift system, but papers by Garfunkel (1988), Steckler and ten Brink (1986), Kohn and Eyal (1981), Omar et al. (1989), Şengör (2001) and Pik et al. (2003) address the topics of erosion and elevation of the African Surface on the flanks of the rifts in this northern part of the rift system.

In the midst of ongoing debate on so-called passive versus active rifting and volcanism at rifted margins, great efforts have been made to establish whether magmatism occurs before, during, or after rifting. It appears that there are no consistent patterns (Menzies et al., 2002). The order in which (1) doming at the Afar, (2) eruption of the Ethiopian traps, and (3) initiation of rifting in the Red Sea, the Gulf of Aden, and Ethiopia took place was reviewed by Şengör (2001) and Burke (1996). Estimates of timing are based on isotopic, magnetic, stratigraphic, and faunal evidence. Review of these data, in combination and with realistic error bars for the different kinds of data, leads to the conclusion that the sequence in which the three events took place is not presently resolvable. Evidence supporting the notion that the pre-eruptive surface in Yemen, which was more than 500 km from the center of the Afar plume eruption at 31 Ma, was above but close to sea level (Menzies et al., 2002) is of limited help in working out what happened where the plume erupted. All three of the phenomena enumerated above appear to have been initiated within a narrow time interval at 31 ± 2 Ma.

An appealing but as yet undemonstrable sequence of events during that interval would be as follows: (1) 1 km or less of doming when the Afar plume-head impinged on the base of the lithosphere; (2) formation of vertical cracks to relieve new stresses set up in response to doming; (3) horizontal propagation of those cracks to places where anomalous stress distribution already obtained: the Levant corner and the southern Ethiopian (then active) volcanic region (Burke, 1996, 2001) along the lengths of what very soon became the Red Sea, Gulf of Aden, and Ethiopian intracontinental rifts; (4) initiation of the eruption of the Ethiopian traps from the Afar plume; (5) dike emplacement from the plume into the cracks on what would become the sites of the three soon-to-be-formed intracontinental rifts; and then, (6) intracontinental rift initiation and rapid extension of several tens of kilometers in the Red Sea, Gulf of Aden, and Ethiopia on the sites of the dike-filled cracks. Ocean floor began to form only much later, beginning between 15 and 10 Ma in the Gulf of Aden and only within the past 5 m.y. in the southern Red Sea.

Burke (1996) attached particular significance to the absence of a laterite-covered African Surface from a roughly circular area ~1000 km in diameter centered on the Afar Dome. He suggested that erosion as the dome started to rise and before the eruption of the Ethiopian traps began could account for the absence of laterite from that roughly circular region. Şengör (2001) disagreed with this suggestion and pointed out that laterites can be preserved on quite steep slopes for relatively long intervals. This may be true in some settings but is not relevant in this case because the lateritized African Surface in this latitude at 30 Ma has been shown to have developed at a low level and to be low lying over a huge area. Interpretative modeling of apatite (U-Th)/He age partial resetting in the Blue Nile canyon (~300 km from the center of the Afar plume eruption site at 31 Ma) led Pik et al. (2003) to argue that erosion began in the canyon as early as 29–25 Ma. The early onset of erosion in the Blue Nile canyon suggests that the elevated plateau topography, which controls most of the present-day Nile hydrology, has existed since the mid-Oligocene (27 ± 2 Ma). Pik et al. concluded that the Ethiopian trap-covered plateau represents a preserved part of a large uplifted dome related to Afar plume impingement and/or to massive underplating triggered by Oligocene continental flood basalt differentiation. Another factor contributing to elevation around the Afar is the presence of the underlying Afar plume tail. It is not presently possible to separate the contributions of these three factors. Our analysis of the absence of laterite in a wider region is consistent both with the results of Pik et al. (2003) and with the view that a kilometer-scale topographic swell was generated in Ethiopia ca. 32–30 Ma, causing laterite stripping on an upwarped African Surface as well as canyon cutting through the rapidly erupted flood basalts and underlying basement.

In the case of the Yemen igneous province, magmatism was also pre-rift, but estimated pre-magmatic dome-related uplift in that area (~500 km from the center of Afar plume eruption at 31 Ma) did not exceed 10–100 m judging from a marine to continental transition in onshore sediment sequences (Geukens, 1966; Menzies et al., 2002).

Summary: Regional Age Brackets for the African Surface

Consideration of the nature of the African Surface, its composite character, and its development in six regions that together cover much of the continent has shown that the surface can be recognized throughout Africa. The African Surface was a product of erosion that began to develop during later Jurassic and Early Cretaceous times (180–120 Ma). It evolved to become a low lying surface with little relief, covering much of Africa and part of Arabia (~33 × 10^6 km^2) by mid-Cretaceous times (ca. 100 Ma). That low lying surface persisted, with various local and regional perturbations, until ca. 30 Ma, when upward flexure of Africa's active swells began. Perturbations to which the African Surface was subjected during the 150–30 Ma interval included episodes of regional flooding by the sea over areas totaling at least as much as 10 × 10^6 km^2, and by tectonic processes, especially long-term intracontinental rift development and deformation driven by convergent processes at Afro-Arabia's Tethyan continental margin. Deformation episodes, particularly that in the Santonian (82–84 Ma) (Guiraud and Bosworth, 1997), involved faulting and folding. The deformation episodes generated high ground, the former existence of which can be best recognized in the folds of the intracontinental rift systems. That high ground was rapidly eroded and its products are recognizable in short-lived pulses of deposition among offshore sedimentary rocks. Although the history of African Surface development has varied from region to region, the coherent character of the land surface and its history has been established. Names, characterizations, and definitions used in research on African surfaces have varied immensely, not only among regions but also through time in the same regions. These variations have put pitfalls in our path and we are certain to have made errors. The following parts of this paper address bauxite occurrence on the African Surface, climatic variation over Africa with time, and the distribution of the African Surface on today's active swells as sources of evidence likely to further improve insights into the African Surface and its evolution through time.

BAUXITE OCCURRENCES AS INDICES OF THE AFRICAN SURFACE: THEIR VALUE AND LIMITATIONS

Bauxites ... and Bauxites

Grandin and Thiry (1983) distinguished (1) lateritic bauxites, which develop on a wide range of parent rocks but appear to have formed worldwide only within narrowly defined geological time slots, from (2) bauxites, which are restricted to specific lithologies with low silica content, high aluminum content, vitreous textures, and high porosities (e.g., limestones, marls, clays, syenites, and basalts). If conditions of high rainfall, good drainage, and moderate to high temperatures obtain, the latter may generate gibbsite-rich regolith (and variable quantities of kaolinite and iron oxides) within a short time. This important distinction explains the existence of bauxite on young geological material under a wide range of humid climates, ranging from atolls in the Solomon Islands, Pleistocene alkaline basalts in Hawaii, Miocene basalts in Oregon, Neogene bauxites of the Caribbean islands, and the Vogelsberg Neogene shield volcano in Germany.

This distinction is echoed in the bauxite classification of Tardy (1993), who, following observations of Boulangé (1984), distinguished *orthobauxites*, which are widespread and in which gibbsite is generated by desilicification of kaolinite in deep tropical weathering mantles (type-1 orthobauxites) or by formation of primary gibbsite directly from some of the minerals of the parent rock (type-2 orthobauxites). Orthobauxites, which involve top-down bauxitization of mature kaolinitic mantles, correspond to the kind of bauxite that we consider here to have covered much of the African Surface in then equatorial Africa during the Late Cretaceous and early Cenozoic. It is helpful, for the purpose of working out the history of erosion surfaces, to distinguish ortho-bauxites from *protobauxites*, which are lateritic soils containing gibbsite only as a minor constituent. Gibbsite may be more abundant at depth in those soils, for example in the young volcanic soils of Mount Cameroon, where internal drainage ensures more intense solute export than nearer the surface. *Metabauxites* are orthobauxites that have been exposed to climates now drier than those at the time of the initial (ortho)bauxitization. This occurs as a consequence of lasting climatic change. Such metabauxites, as occur for instance in southern Mali, contain boehmite and exhibit a whitish appearance. Finally, *cryptobauxites* evolve in humid equatorial forest zones under a cover of alluvial or colluvial material.

The nature of the parent rocks, particularly the iron oxide mineral and quartz content, adds variety to bauxites. If the Fe/Al ratio of the parent material exceeds 1, as on Precambrian banded iron formations, Tardy (1993) proposes that the resulting alumino-ferruginous products be called *conakrytes*, after a type locality in the Conakry area of Guinea, rather than bauxites. Conakrytes are therefore identical to orthobauxites in climatic terms but develop on rocks in which Fe/Al >1. The recognition of different varieties of bauxite leads us to suggest that in regions in which the African Surface has been identified—either in this study or in earlier studies—but in which orthobauxite has not been observed (e.g., Uganda, and eastern, central, and southern Afro-Arabia generally), ferruginous laterites may be protobauxites or conakrytes formed during the Late Cretaceous to Paleocene interval (ca. 90–55 Ma).

On quartz-poor parent rocks, gibbsite forms readily and may lead to rapid bauxite formation, whereas on quartz-rich rocks, the formation of kaolinite inhibits the direct formation of gibbsite, so that orthobauxite development on quartz-rich lithologies requires one or both conditions of (1) longer periods of geologic time and (2) exceptionally humid climate.

Examples of Orthobauxites and Laterites in Afro-Arabia Inherited from the Last Great Bauxite-Forming Interval, 70–40 Ma

Widespread orthobauxite formation has occurred only four times during Earth's Phanerozoic history. A mainly Paleogene bauxite-forming event, between ca. 70 Ma and ca. 40 Ma was the most recent. A postulated Neogene episode based on evidence from karst occurrences in southeast Europe remains poorly constrained. Bauxite developed between the Campanian (70 Ma) and the late Eocene (34 Ma) over a region extending from southeast Spain and Turkey through Hungary, as well as in Senegal and Guinea. This area embraces ~30 degrees of both latitude and longitude. Tardy and Roquin (1998) distinguished orthobauxites in the western part of this area from protobauxites in the east of Africa. The two kinds of weathering occurred in different areas on the continent-wide African Surface. The difference can be attributed to counterclockwise rotation of Africa between the Jurassic and the present about a poorly defined pole in the Atlantic (Fig. 18) (Scotese and Sager, 1988). Conditions are also likely to have been more humid in West Africa from 180 Ma onward because zonal winds that had crossed the central Atlantic Ocean would have penetrated that area with more moisture than they carried when they reached the eastern parts of Afro-Arabia. The west of the continent remained within orthobauxite-forming equatorial climates while the eastern part, which rotated farther latitudinally between the Jurassic and the Oligocene (200–30 Ma), experienced somewhat drier climates, in which ferruginous laterite formed with gibbsite as a minor constituent. That is, protobauxites were the main weathering products.

Staircases of Bauxite- and Laterite-Capped Surfaces in Their Key Area: West Africa

The idea that bauxites and laterites were forming episodically in some areas and continuously in others between ca. 100 Ma and ca. 40 Ma has led to the notion that bauxites and laterites formed by deep weathering on the Gondwana continents in an age-related sequence, with the thickest developments of weathered rock and most, if not all, the bauxite deposits capping the highest topographic surface in any one region (see, e.g., Grandin 1976; Aleva 1994: p. 43–44; Boulangé and Millot, 1988; Gunnell, 2003). The sequence has been geochemically fingerprinted (Table 1, Fig. 19) and supports the idea of a chronosequence of pediment surfaces each sealed by lateritic crusts with distinctive geochemical and petrographic characteristics (Boulangé et al., 1973).

The staircase of lateritic surfaces is most simply explained by a model in which the bauxitic African Surface occupied low ground until 30 Ma when the basin-and-swell structure of Africa began to develop. In that case, all the lower steps, or pediments, of the staircase (1) formed on the flanks of the rising swells, (2) are lateritic rather than bauxitic because they formed after the bauxite-forming interval had ended, and (3) contain detrital fragments of lateritic and bauxitic crust eroded from higher levels. Lower-level surfaces could be expected to contain fragments from more than one higher level. Gunnell (2003, Fig. 2 therein) illustrated what may be a more complicated situation. The three highest surfaces in the Fouta Djallon are all underlain by bauxitic weathered profiles. Michel (1973), in a strictly Davisian approach, regarded the two higher surfaces (Labé and Dongol Sigon) as being chronologically and genetically distinct from one another, both being older than the African Surface. Because the ongoing uplift of the seismically active Fouta Djallon, like that of all of Africa's other swells, was initiated at 30 Ma, an alternative possibility is that all the surfaces underlain by bauxitic weathering are representatives of the composite African Surface, with bauxitic weathering having developed before 40 Ma. The surfaces, in that case, record

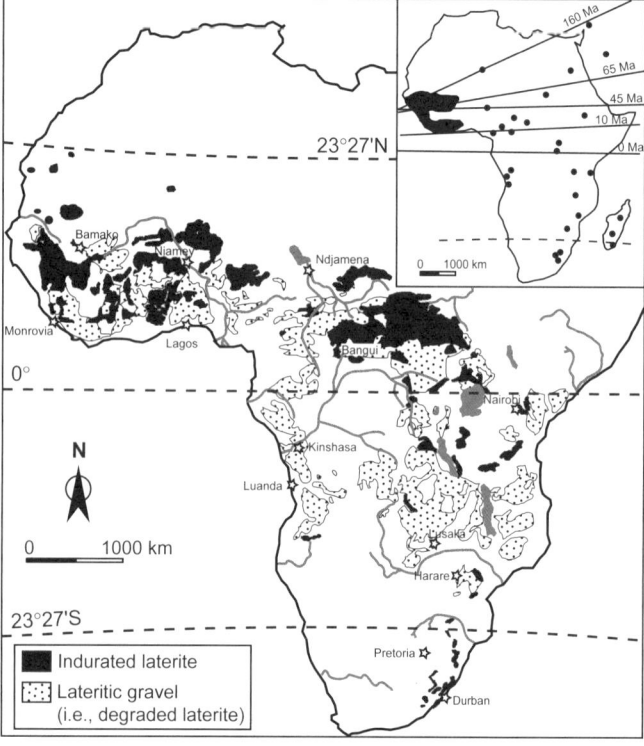

Figure 18. Distribution of laterites, bauxites, and lateritic soils in Africa. Inset panel shows highest geographic concentration of bauxite occurrences in Africa and approximate migration of geographical equator as a result of the counterclockwise rotation of Africa about a pole situated in the Atlantic Ocean (shown here as west of Liberia). West Africa has remained close to the equator ever since the Karroo-PIPPE episode with which the history of the African Surface began at 184 Ma. This explains the potential for continuous bauxite generation in farthest west Africa during that time, including conakrytes (see text for definition) near sea level in recent times. By contrast, across the rest of Africa, bauxites are scattered and localized, and may not strictly represent orthobauxites (see text for definition). Sources: Boulet et al., 1971; FAO-UNESCO, 1976; Petit, 1982, 1985; Patterson et al., 1986; Boeglin, 1990; McFarlane, 1991; Boulvert, 1996, 2003; Tardy and Roquin, 1998.

TABLE 1. GEOCHEMICAL CHARACTERISTICS OF LATERITIC ROCKS CAPPING ON FIVE DIFFERENT GEOMORPHIC SURFACES RECOGNIZED ON THE SWELLS OF BURKINA FASO AND CÔTE D'IVOIRE

Lateritic seals	No of samples	Relative height above local base levels (m)	Average depth to bedrock (m)	Ironstone thickness (m)	SiO_2 (mean % ± 1σ)	Al_2O_3 (mean % ± 1σ)	Fe_2O_3 (mean % ± 1σ)
African surf. (Bauxite)	45	150–500	60	≤ 20	3 ± 2	51.4 ± 10	18 ± 13.5
Relief Intermédiaire	83	80–170	30	6–10	13.8 ± 4.6	16 ± 5	57.7 ± 8.5
Haut-Glacis	40	50–100	20	5–10	28.6 ± 8.2	17.6 ± 3	43 ± 5.8
Moyen-Glacis	13	10–30	10	1–5	38.6 ± 12.6	16 ± 5	35 ± 6.3
Bas-Glacis	7	2–10	5	0–2	41 ± 8	11.5 ± 4.7	33.5 ± 12.2

*Compiled after Boulangé et al. (1973); Grandin (1976); Tardy and Roquin (1998).

episodic uplift during the past 30 m.y. In Blafo-Gueto, in south-central Côte d'Ivoire, a single major bauxitic horizon occupies plateau summits (Grandin, 1976) (Fig. 14). Deep weathering extends below the plateau surface in Blafo-Gueto for tens of meters. The plateau surface is here suggested to be the African Surface and the weathering is considered likely to have extended over the entire interval defined by the marine intercalations of the Iullemeden Basin (ca. 100–40 Ma). Grandin (1976) and Colin et al. (2005) assigned an Eocene (53–34 Ma) age to the surface that we call the African Surface at Blafo-Gueto. This accords with results from the Iullemeden Basin, where rocks as young as Lutetian (ca. 45 Ma) are both lateritic ("ferrallitic" in the French literature) and flexed upward on the southwestern flank of the

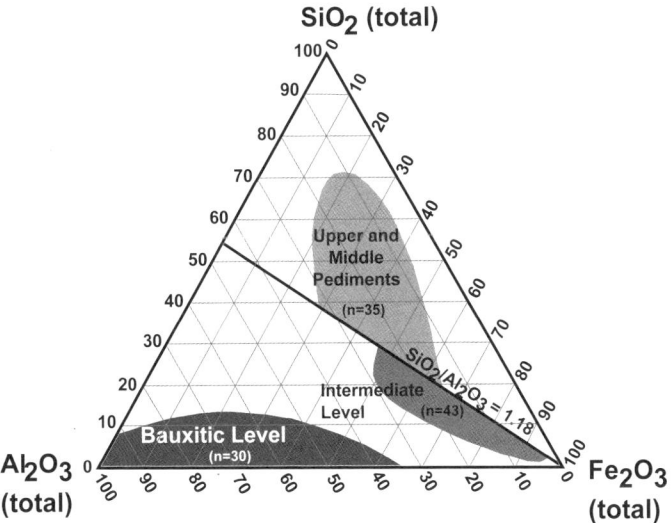

Figure 19. Distinction between successive generations of West African laterites based on geochemical fingerprinting (after Boulangé et al., 1973). Rocks of the bauxitic level formed on the African Surface when it was low lying between ca. 100 Ma and 30 Ma. Since the African Surface began to be uplifted on swells at 30 Ma, intermediate-level laterites and those of the upper and middle pediments have formed on the lower slopes of the swells. According to petrographic distinctions made by Grandin (1976), they commonly include pebbles derived from the laterites and bauxites of the African Surface.

Aïr swell, one of Africa's 30-Ma-and-younger swells (Fig. 20). Later, nonmarine sediments that have only been involved in the flexure of the swell to a lesser extent, or not at all, were assigned by Boudouresque et al. to the *continental terminal*.

In the Iullemeden Basin (Fig. 9) the intercalation of fossiliferous marine horizons has permitted the identification of three episodes of alteration (Boudouresque et al., 1982), described as *ferrallitique*, that led to the formation of "bauxitic levels." Those episodes were during (1) late Turonian to Santonian times (ca. 90–84 Ma); (2) later Campanian times (ca. 75–70 Ma); and, (3) perhaps with varying intensity over a longer interval extending from latest Maastrichtian times (ca. 66 Ma) throughout the Paleocene (65–53 Ma) and early Eocene (53–45 Ma), ending in Lutetian times (ca. 45 Ma). The local record in the Iullemeden Basin (Kogbe, 1981; Boudouresque, et al. 1982) and the consensus among students of the worldwide record (e.g., Thomas, 1994; Aleva, 1994; Tardy and Roquin, 1998) recognize deep lateritic and bauxitic weathering as having occurred between ca. 90 Ma and ca. 40 Ma, and therefore during a large fraction of the interval when the African Surface was developing. In some places, as in the Iullemeden Basin, discrete episodes within the long interval can be defined. The most intense deep weathering is considered to have occurred during the Paleocene (Schmitt, 1999; Zachos et al., 2001; Colin et al., 2005; and see later section on paleoclimate below).

The *continental terminal* in northwest Nigeria is represented by the Gwandu Formation of Eocene to Miocene age (Kogbe, 1981), which lies unconformably over all the older beds of the Iullemeden Basin. It consists of red and mottled kaolinitic clays with sandstone intercalations and reaches maximum thicknesses of ~350 m. The Gwandu formation and equivalents cover 22,000 km² in northwest Nigeria and 150,000 km² in southwest Niger, where thicknesses attain maxima of 450 m. It also extends into southeast Burkina Faso and Mali, and covers vast expanses of eastern Niger and much (possibly up to 75,000 km²) of the Chad Basin. There it laps onto the Precambrian basement in the south and northeast of Lake Chad, onto Cretaceous sediments in the west, and onto Mesozoic or Paleozoic sandstones to the north. Everywhere it consists of the products of denudation of the

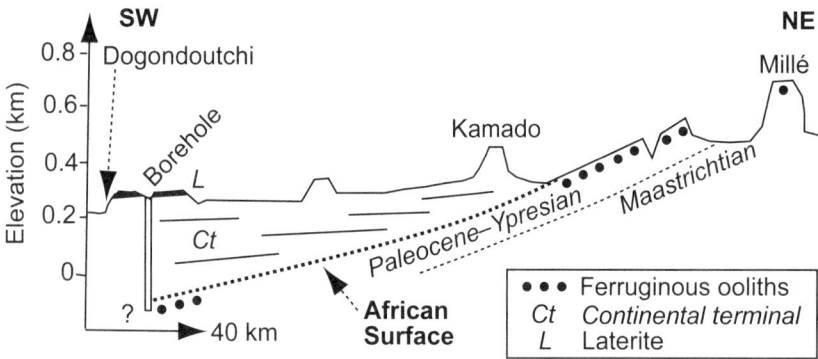

Figure 20. Sketch cross section indicating *continental terminal* (Ct) deposits of post Ypresian (ca. 50 Ma) but probably <30 Ma age, lying unconformably on the African Surface, between Dogondoutchi and Millé in southwest Niger. These are ~200-m-thick deposits in this part of the Iullemeden Basin. The unconformity results from the Aïr swell to the northeast of the figure (see Fig. 2 at ~10°E) having become elevated during the past 30 m.y. Modified after Boudouresque et al. (1982).

lateritic and bauxitic cappings that developed during the Paleocene and early Eocene, but also during Cretaceous times.

The Gwandu Formation and its correlatives overlie the older strata with an angular disconformity because those Cretaceous and older Cenozoic units that had been deposited earlier on the African Surface had acquired a dip due to swell-related uplift in the regions surrounding the modern basin (Fig. 20). The angular unconformity indicates that the beginning of crustal uplift, of which continuing developments are responsible for the dip of the Gwandu Formation itself, took place before the Gwandu Formation began to be deposited. That uplift made it possible for the deeply weathered lateritic material to be eroded. Erosion and uplift occurred between the late Eocene (ca. 36 Ma) and the early Oligocene (34–28 Ma). Lateritic horizons (there are no bauxites reported within the *continental terminal* of the Iullemeden Basin) are similar to those mapped at lower elevations in Blafo-Gueto. In both areas, those gravels contain laterite fragments that are products of erosion from higher and older horizons including those of the African Surface (Grandin, 1976; Colin et al., 2005).

Climatic change following establishment of the Antarctic ice sheet ca. 34 Ma, increased continental freeboard of Africa as a result of ~50 m of sea-level lowering. Combined with the initiation of tectonic basin-and-swell formation ca. 31 Ma, these two events together marked the termination of African Surface development and brought the great bauxite-forming interval to a close. There have been several later episodes of laterite formation in Afro-Arabia. None, however, have been as intense or as protracted as the bauxite forming episode that, as evidence from the Iullemeden Basin indicates, extended from ca. 100 Ma to late Eocene times. The later episodes typically (1) occur on lower hilltops and hill slopes (as in the Central African Republic; see earlier section), (2) incorporate detrital fragments from older laterites, and (3) contain no bauxites.

A synthesis of bauxite occurrences across Africa can be attempted, following a compilation of French- and English-language reports and an existing synthesis by Petit (1985). In southern Africa, the better known bauxite occurrences are reported from Malawi on the Lichenya Plateau, near the Mozambique border, and in Zimbabwe and Mozambique on the crest of the Great Escarpment, on either side of the border in the Penhalonga area. In South Africa, bauxites occur in the currently wettest and most elevated regions of the Natal Drakensberg. Considering their absence from areas farther to the west, it seems likely that the current climatic gradient has changed little since the East Antarctic ice sheet formed, establishing modern atmospheric circulation patterns and corresponding humidity gradients after ca. 34 Ma.

In central Africa bauxite occurrences are poorly known and probably not widespread. They have been described in the Dondo and Luanda areas of Angola, and in the Mayombe uplift of the northwestern Democratic Republic of Congo (Vanderstappen and Cornil, 1955; Stas, 1959). Ferruginous laterites are more widespread. Cameroon potentially constitutes a key region in working out the relationships between the aluminous and ferruginous laterites of central Africa, although Neogene tectonism makes it more complex than in neighboring sub-Saharan regions (Morin, 1989; Burke, 2001). Bauxites in north-central Cameroon capping the Minim-Martap surface at an elevation of ~1.2 km a.s.l. are developed on basalts of poorly defined age. Whether these are orthobauxites or simply protobauxites or conakrytes (see terminology above) is debatable until further research is carried out. The highest surface of Cameroon and Nigeria, the "Grassfields surface," is overlain by Miocene basalts, younger volcanic rocks than those on the lower-lying Bamileke Plateau, which carries volcanic rocks of suspected Eocene age (Fritsch, 1978). The lower Meiganga-Bamoun surface carries ferruginous laterites that, as in the case of the intermediate level in West Africa, are reported to contain bauxitic debris. The Meiganga-Bamoun surface could therefore be an equivalent of the intermediate level of West Africa, although the laterites have not been studied in sufficient detail to confirm the correlation. On the basis of elevation criteria and equally imprecise petrographic comparisons of lateritic caprock, the Meiganga-Bamoun surface can be discerned continuing eastward into the Central African Republic.

Caution, however, is urged with regard to correlating the bauxites of Cameroon with the African Surface in the systematic way achieved by Michel (1973) and Grandin (1976) in sub-Saharan West Africa. Elevations in the Cameroon region are no sure criterion for continent-wide correlations because of the complex block-faulted geological structure (Morin, 1989) and a tectonic history linked to post–30 Ma volcanism (Burke, 2001). Igneous activity began along the Cameroon line in the Paleocene at 65 Ma, with the first eruptions of the *granites ultimes* (Burke, 2001). There was a change to dominantly basaltic volcanism at 30 Ma, and basaltic activity has persisted both on the individual relatively small swells of the Cameroon Line and on other larger swells such as Ngaoundere ever since (Burke, 2001). It seems that humid climates existed from Paleocene to Miocene times in this region, which still records the highest extremes of rainfall of West and central Africa.

In terms of climatic change, Cameroon experienced a shift from drier to more humid climates from the earlier Cenozoic while the rest of West Africa went through the reverse process. The high rainfall totals of Cameroon therefore make it a possible exception within Africa. Bauxite development, possibly protobauxite formation, may have continued into the Neogene at topographically suitable sites and on suitable lithologies. The reported presence of bauxite on Miocene flows covering the Bamileke surface, but also on the more elevated "Grassfields" plateaus in Cameroon (Morin, 1989), for instance, is therefore neither a sufficient criterion to doubt the radiometric ages of the flows nor a sufficiently strong case to doubt any correlation of the sub-volcanic Bamileke surface with the African Surface. Both Fritsch (1978) and Morin (1989) concluded that the Minim-Martap/Bamileke surface was Eocene, which is consistent with our terminal age range for the African Surface.

One conclusion from the interpretation of bauxite occurrences in Afro-Arabia is that it is important to attempt to discriminate between bauxites inherited from the Late Paleocene Thermal Maximum (see Early Cenozoic Times, below) and younger or older occurrences that do not carry the same global significance. Considering the diversity of the bauxitic materials discussed here, their diverse climatic significance, the current difficulties that exist in dating bauxite, and the intense neotectonic movements in Cameroon that are likely to have fragmented and offset the African Surface compared with neighboring regions of West and Central Africa, Cameroon will not be the best suited area to reconstruct the African Surface until more work is carried out.

French researchers working in the Congo Basin have analyzed the geochemical destruction of ancient ferruginous laterites by climatic rehumidification, involving gibbsite and goethite formation from kaolinite and haematite, respectively. This has been studied in detail in the SE Central African Republic on the Haut-Mbomou surface at the edge of the Congo Basin (Beauvais and Roquin, 1996), but the homogeneity of the soil cover across into SE Cameroon, Gabon, Congo, and the northern Democratic Republic of Congo (Ségalen, 1995) indicates a widespread impact of this process in direct relation to the re-expansion of wet evergreen forests out of the Congo Basin, at times, thus far undated, when climates became wetter. Although increasingly well understood, the process of degradation of older ferruginous laterites into gibbsite-rich soils, often also characterized by a gravel horizon, has been crucial in miscorrelating these pseudobauxitic soils with degraded forms on the African Surface where it has been downwarped into the Congo Basin.

In the Central African Republic, which, with Burkina Faso, is the most extensively lateritized nation in the world, laterites were exhaustively surveyed by Boulvert (1996). They are nowhere bauxitic save a small outcrop in the northeast, on the Wadda Plateau where the bedrock consists of Cretaceous sandstones. Between the more complex areas of tectonically disturbed Cameroon and the East African rift system, tiered generations of laterites are clearly identifiable in the topography of the Central African Republic, but the range is not as complete as on the Fouta Djallon and Guinea swells of West Africa. Petrographic and geochemical studies of deep weathering profiles on poorly differentiated topographic levels, at the local scale of single interfluves in the Central African Republic, have weathering characteristics akin to weathered mantles that, in comparable environments in West Africa, are specific to defined geomorphic levels (Boulvert, 1996). Kilometer-scale toposequences display mature laterites (no quartz, abundant hematite) capping mesas above widespread levels of less mature higher pediment facies (quartz, goethite, rare gibbsite) and, farther downslope, Lower Pediment hardpan caused by active seasonal waterlogging. These weathering signatures, which reflect relative age ranking in the topography, are considered to indicate a geomorphic evolution similar to that in West Africa but less well expressed topographically as a result of comparatively limited uplift in the Central African Republic.

The *surface centrafricaine* is covered by laterite exhibiting the higher pediment facies: if correlated with the widespread higher pediment occurrences of the West African craton (e.g., most of central Burkina Faso and northern Côte d'Ivoire: see Eschenbrenner and Grandin, 1970; Eschenbrenner et al., 1974; Eschenbrenner and Badarello, 1978), the *surface centrafricaine* should therefore be of late Miocene to Pliocene terminal age. The Chad and Congo Basins on either side of the Central African Republic swell became depositional basins for the *sables rouges* facies of the *continental terminal* on the Chad side, and for the *sables ocres* on the Congo side.

Bauxites in Africa That May Have Formed since the End of the Main Bauxite-Forming Interval, i.e., after 40 Ma

Bauxite deposits are rare in East Africa. Observing the coincidence between elevated rift-flank uplifts and the few bauxites and red latosols that have been suggested to result from geochemical destruction of former laterites (Beauvais and Roquin, 1996), Tardy and Roquin (1998) proposed that East African bauxites, unlike those of the rest of Africa, were not inherited from the last bauxite-forming maximum. They suggest that they were, instead, formed during Neogene times. The rationale for this idea was that high temperatures, even associated with high rainfall, favor ferruginous lateritization rather than bauxitization of the regolith. It is only as a result of the cooler temperatures afforded by higher elevations resulting from surface uplift, coupled with the high rainfall that characterizes tropical highlands, that bauxite could develop in topographically suitable sites within a continuous red latosol cover. This would have happened because the latosols themselves were generated by a softening of earlier Cenozoic ironstone cappings as a consequence of wetter climatic conditions establishing themselves under an expanding evergreen forest cover. Their source rocks were already heavily depleted in quartz as a result of past intense weathering, although the latosols retained sufficient kaolinite for gibbsite to form.

The core message of this interpretation is that bauxite formation is not only rainfall dependent, but also dependent on the initial silica content of the parent material subjected to intense lateritization or, as in this case, repeated lateritization. Because of these considerations, Tardy and Roquin (1998) argued that the surface, here interpreted as representing the African Surface in East Africa, was not originally bauxitic but became locally bauxitized in Neogene times as a consequence of uplift. This speculative assertion can, however, be overturned: it could be suggested that, instead, a bauxite-capped African Surface has been stripped from many places because the rates of uplift and denudation close to the rifts are much greater in East Africa. This is also likely for Madagascar, where bauxites and laterites are rare in spite of suitable long-term climatic conditions. Growing evidence now indicates that uplift of the active rifts of East Africa is very recent, i.e., younger than 15 Ma along the shoulders (Burke, 1996, p. 383) and with a peak of denudation between 7 and 2 Ma determined by apatite helium thermochronology (Spiegel et al., 2007). It is therefore unlikely that bauxite developed in such a short time span, and more likely that the older regolith was stripped by the late Neogene denudation, with remnants being well preserved only on continental divides spared by the denudation. This may be true in the Western Rift region, which was uplifted latest (5–2 Ma, see Spiegel et al., 2007) and where highly aluminous highland soils have been reported on the elevated surfaces of Rwanda, Burundi, the Byomba, and Butare surfaces of Rossi (1980), and the Congo-Nile watershed more generally (Kellog and Davol, 1949; Ruhe, 1956; Van Wambecke, 1963; Jongen et al., 1970; Sys, 1972; Frankart, 1983; Mutwewingabo, 1989). The ages and paleoenvironmental implications of these soils are only beginning to be discussed in the literature (Caner and Bourgeon, 2001). They may hold more revelations.

Synthesis on Bauxites and Laterites

Geomorphologists of the mid–twentieth century, particularly in West Africa, tended to consider that erosion cycles were primarily climate driven. For example, authors such as Brückner (1955) and Michel (1973) suggested that the lateritic staircases of West Africa formed as a result of Quaternary pluvial and interpluvial cycles. Brückner, in particular, was influenced by the ideas that his uncle had helped to develop for the Quaternary history of the Alps (Penck and Brückner, 1901–1909), and the French climatic geomorphology professed in Strasbourg (where Michel was based) has historically overlapped with the German outlook. Those ideas led to a perception that erosion surfaces could be defined by a particular climatic signature, by a set of climatically defined geomorphic processes, and by weathering mantles and the topographic surfaces on which they formed, which were regarded as being coeval and cogenetic. In West Africa, the African Surface equivalent was considered to be bauxitic and any tract of the African Surface established on the basis of elevation criteria but not associated with a bauxite deposit was suspect.

Climatic geomorphologists were understandably tempted to associate the different generations of West African lateritic pediments with the Quaternary because they recognized that interval in earth history as one most influenced by sharp climatic oscillations. That led to a second and equally misleading perception, namely a significant underestimate of the time required for the formation of laterite by approximately one order of magnitude (Tardy and Roquin, 1998; Colin et al., 2005), and therefore to an erroneous sense of continental denudation chronology. The position in this paper is that erosion surfaces generally, and the African Surface in particular, are predominantly gravity-driven features of the landscape. Tectonic and eustatic changes in base level, in addition to time, are the key factors in denudation. Climatic factors, within the frost-free intertropical zone, arguably have a more limited impact at the time scale of geologic eras.

The diversity of ferruginous and bauxitic regolith characteristics across the African Surface in the four corners of Africa is consistent with the interpretation of the African Surface as a composite land surface. Although the African Surface is probably systematically wedded to the presence of bauxite in West Africa, the scarcity of bauxite elsewhere on the African continent does not invalidate the recognition of the widespread occurrence of the African Surface outside West Africa's boundaries. By pointing out that there are several possible geochemical and geomorphological pathways to bauxite formation, Tardy and Roquin (1998) warned usefully against one potential pitfall in using the African Surface as a morphostratigraphic datum: bauxites are not all necessarily a legacy of processes that culminated at the time of the Late Paleocene Thermal Maximum. Tardy and Roquin also hinted that the African Surface need not systematically be clad in bauxite outside West Africa. The African Surface in central and eastern Africa is predominantly covered by ferruginous laterite as, for example, on the Buganda surface. It is of crucial importance to be cautious in assuming (1) that bauxitic outliers are associated with remnants of the African Surface, (2) that those bauxites are necessarily legacies of the 70–40 Ma bauxite-forming maximum, and (3) that bauxite is the ubiquitous stamp of the continent-wide African Surface. Nevertheless, using the African Surface as a reference benchmark to constrain post-Eocene surface uplift retains its value as a geomorphic tool for estimating magnitudes of continental-scale crustal deformation, and as a complement to other deformation indicators such as, for instance, paleoshoreline marker rocks as used by Sahagian (1988).

The acceptance among laterite mineralogists that gibbsite and goethite are hydrated minerals related to humid and cooler climates, whereas boehmite and hematite are dehydrated minerals that are the product of warmer and drier tropical climates

(Tardy, 1993), constitutes a helpful guide to understanding the lateritic formations that cap the African Surface across the African continent. As physically resistant and long lasting features of the tropical landscape, laterites are subject to internal mineralogical readjustments that reflect the effects of regional climatic changes on the parent lateritic and bauxitic material inherited from the 70–40 Ma bauxite-forming maximum. The record of these climatic changes, although informative, is subordinate for the purpose of this paper to the recognition of the African Surface as a polygenetic land surface formed during ~150 m.y. of denudational and bioclimatic evolution.

THE IMPRINT OF CONTINENTAL-SCALE PALEOCLIMATIC VARIATIONS IN AFRO-ARABIA

Cretaceous Times, 140–65 Ma

The mid-Cretaceous interval (ca. 110–90 Ma) was a time when continental ice sheets were absent on the Earth: it was a "hothouse" Earth. It has been considered "the warmest interval during the Phanerozoic, and late Cretaceous climates were also quite warm but experienced a series of fluctuations to cooler temperatures. The mid-Cretaceous record provides abundant evidence of humid continental climates and reduced areas of continental aridity" (Fawcett and Barron, 1998, p. 29; see also Huber et al., 2003). Conditions in the parts of Africa that lay in the humid Tropics during the Cretaceous would have been highly favorable to deep weathering, and the heavy rainfall associated with the Intertropical Convergence Zone would have fostered much erosion. The late Aptian (ca. 112 Ma) evaporites of the northern South Atlantic, Africa's only thick Cretaceous evaporite deposits (Burke, 1975, 1977), are presently distributed between the equator and 15°S. They give an indication of where the southern desert latitudes of the time lay. Aptian latitudes of ~10–25°S occupied a region presently 0–15°S. This is consistent with the latitudinal position of Africa based on plate reconstructions. The equator, and by inference the convergence zone with the optimum conditions for heavy rainfall, intense erosion, and deep weathering, lay over parts of Africa that are now farther north than they were during Cretaceous times. As the continent rotated slowly counterclockwise, the equator and those equatorial conditions migrated. Because the pole describing this motion was in the Atlantic, latitudinal migration was much less in Guinea than in the Horn of Africa. Greatest Cretaceous weathering is likely to have occupied a zone straddling a line drawn roughly from Conakry to Aden. Since ca. 180 Ma the Bulge of Africa has lain adjacent to the waters of the Central Atlantic Ocean and has strongly felt the influence of moisture-laden zonal winds through much of that time. The present condition, in which the NE trade winds dominate on the Bulge, is not typical of the previous 180 m.y.

Global terrestrial climates and vegetation distributions for Maastrichtian times (71–65 Ma) have been illustrated by Upchurch et al. (1999), who tested five environmental models (Fig. 21) controlled by fossil plant distributions and continental positions reconstructed on the basis of ocean-floor magnetic anomalies. Afro-Arabia, with seven fossil plant localities, was dominated from the Bulge of Africa to what was then ~20°S on the east coast by "tropical semideciduous forest" with a limited "tropical rainforest" concentration, which varied in extent depending on proposed models, close to the then equator. A relatively small area of Northern Hemisphere desert and semidesert occupies northwest Africa in all five models, and a more extensive region of desert and semidesert adjacent to Africa's west coast in the South Atlantic Ocean was centered on the area of the present Congo Basin. The models also vary in the extent of "subtropical broadleaved evergreen forest and woodland" to the east of the southern desert area. All the models but one characterized the southernmost tip of Africa as occupied by "temperate broadleaved and coniferous forest."

The major purpose of these new models is to reconcile previous global circulation models (GCMs), which were essentially based on marine data, with the terrestrial record. Equable climates in large continental interiors caused by low meridional temperature gradients could not be reproduced by earlier GCMs. Based on averaged annual data that did not incorporate seasonal contrasts, these tended to predict cold continental interiors, so that the equable nature of climates in the Cretaceous hothouse Earth was called into question (DeConto et al., 1999). Land cover parameters, and particularly vegetation masses in continental interiors acting as major engines of feedback on global climate systems, appear to generate better fits with evidence from the geologic record on large continents such as Africa.

The inference of a desert climate in the area of the present Congo Basin appears somewhat more extreme than the paleoenvironment proposed by Axelrod and Raven (1978): their biogeographic reconstructions for Africa during the Late Cretaceous and Paleocene are identical but place, within the desertic semicircular area of west-central Africa, a savanna woodland vegetation with sclerophyll woodland enclaves rather than open desert. The difference, however, is one of drought intensity rather than any major paleogeographic discrepancy in the distribution of bioclimates. The geologic record identifies eolian sandstones in the Congo Basin during Paleogene times (*grès polymorphes* enclosing silcrete levels). These represent reworked Cretaceous Kwango series sandstones, in which the feldspars have been weathered into clay. Toward the southern edge of the basin (in Shaba), where sand grains clearly record eolian transport and desert varnish (Mortelmans, 1964; Claeys, 1947; Chevallier et al., 1972), a gradient of increasing aridity has also been recognized. These environments have been compared with some parts of the current Kalahari Desert, which receive sufficient moisture to dissolve silica, generate silcrete, and rapidly fossilize fragile freshwater shells before they are damaged.

Three implications of the models for Africa, of which Figure 21 is a typical example, emerge: (1) Laterite-forming conditions may not have extended very far into southern Africa during the peak of the hothouse period. The most southerly deeply weathered lateritic extension of the African Surface may have

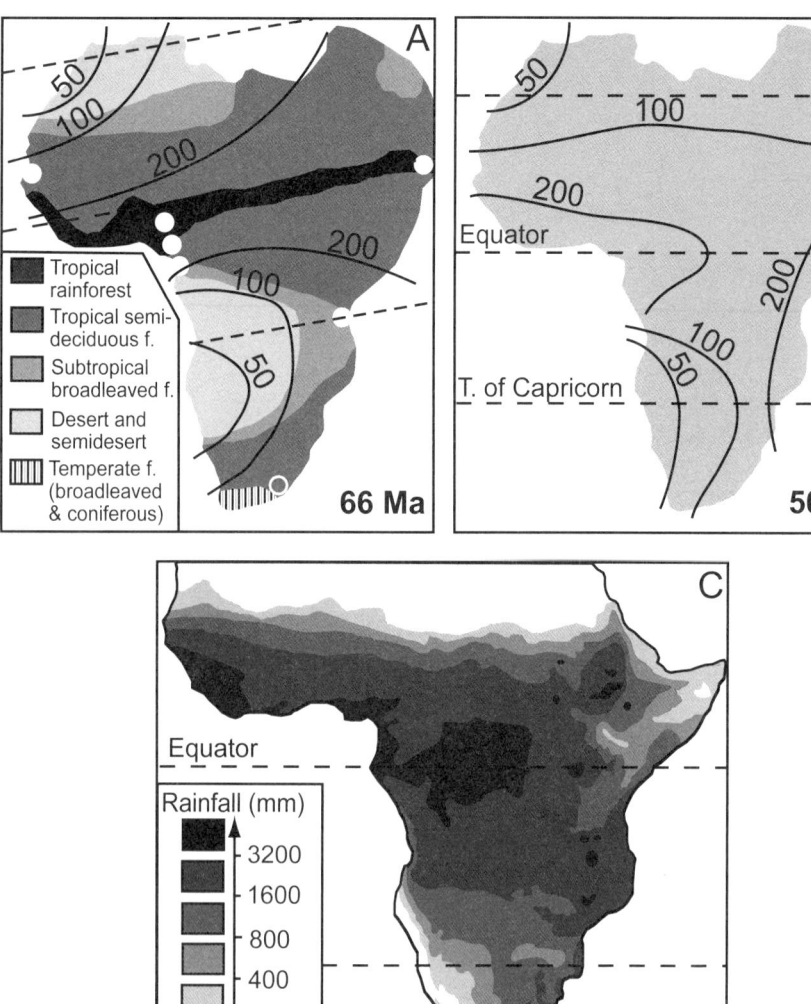

Figure 21. Africa's bioclimate since Late Cretaceous times. (A) Model of Maastrichtian (ca. 66 Ma) biomes across the African continent calibrated against lithologic and paleontologic primary data points (after Upchurch et al., 1999). White dots—evidence of subtropical evergreen forest or woodland based on leaf assemblages, fossil tree stumps, and a minimum of two to three pollen types assignable to tree genera currently in existence. White circle: evidence of tropical rainforest based on the occurrence of equatorial coals in the sedimentary record. The major feature is the presence of a desert in what now corresponds to the Congo Basin and Angolan swell. Given that the Antarctic ice sheet was not yet in existence, this desert was unrelated to an ancestor of the Benguela Current. The subtropical African desert during hothouse Earth was essentially a zonal desert, a Southern Hemispheric Sahara with a smaller symmetric occurrence in the Maghreb (Giresse, 1982). Continuous lines—distribution of isohyets during Late Cretaceous times according to Parrish et al. (1982). Values are not given in absolute (e.g., millimeters) but in relative terms, with areas >200 corresponding to the most humid, and areas <50 corresponding to desert conditions. Dashed lines—paleo-equator and Tropics. (B) Distribution of isohyets in Middle Eocene times according to Parrish et al. (1982). Values as in panel A. (C) Rainfall distribution (in mm) across Africa today.

been the crest of the Zambian swell ca. 71–65 Ma, but duricrusts were probably not lateritic until Paleocene times (ca. 65–53 Ma) when counterclockwise rotation had carried southern Africa closer to the paleo-equator. (2) As a result of climatic conditions, denudation rates in the two subtropical deserts were probably low on either side of the equator, i.e., in northwest Africa and the Congo-Angola-Gabon region (Fig. 21). (3) Finally, most other parts of Africa appear to have been forested, and therefore favorable to deep weathering with low rates of physical erosion.

It seems, therefore, that most if not all of Afro-Arabia during hothouse Earth times was denuded to a low elevation linked to high sea levels (for example the trans-Saharan seaway through the Iullemeden Basin and the Benue Rift system), low tectonic activity, and extreme climates (hot and humid or hot and dry). All of these factors combined to reduce continental relief and to dampen the effects of erosional processes. That time interval corresponds to the period when the African Surface acquired its identity as a continent-wide topographic feature of low mean relief and low elevation—except for a few scattered, lithologically controlled landforms and the resistant valley-and-ridge stumps of some Mesozoic rift shoulders maintained proud of the main land surface.

Earlier Cenozoic Times, 65–34 Ma

During this interval, climate oscillated and generally began to cool with a short-lived (110–220 k.y.) very warm episode peaking ca. 53 Ma. Until recently called the Late Paleocene Thermal Maximum, this paleotemperature spike is now termed the Paleocene-Eocene Thermal Maximum (PETM) or Early Eocene Thermal Maximum. It has become the principal criterion used by the International Subcommission on Paleogene stratigraphy to recognize the Paleocene-Eocene boundary. This warm episode was initially identified in deep-sea sediments (see Zachos et al., 2001), but the related carbon isotope excursion has also been recognized on land and matches the timing of the Paleocene-Eocene boundary based independently on continental vertebrate stratigraphy (Huber et al., 2003).

The PETM seems to have left its mark on the African Surface in the form of a peak in lateritic and bauxitic evolution in West Africa (see, e.g., the ^{40}Ar/^{39}Ar age for Mn-rich laterites capping the African Surface reported in Burkina Faso [Colin et al., 2005]), but also possibly in South Africa, where preliminary ^{40}Ar/^{39}Ar ages (obtained on todorokite for an Mn-ore enrichment event in saprolite capping the African Surface) extend back to ca. 43 Ma (Gutzmer and Beukes, 2000). This setting may provide older ages as work progresses to other sites of the South African manganese fields and uses target-minerals with a more argon-retentive tunnel structure (cryptomelane rather than todorokite). The PETM is also very well defined by palynozones and the nature of lacustrine deposits in the Cenozoic deposits of the Iullemeden Basin (Boudouresque et al., 1982). During Paleocene times, marine sedimentation in Niger was dominated by limestone, phosphate, and palygorskite, and was contemporaneous with deep kaolinitic weathering in adjacent land areas. The *continental terminal* represents the products of erosion of these deeply weathered mantles during the subsequent phase of swell updoming, which coincided with a change of climate toward drier conditions—testified to by savanna floristic assemblages appearing in the Middle Eocene palynozone of the Iullemeden Basin. After the PETM, and until the climatic and tectonic changes of the end of the Eocene, the African Surface is envisaged to have continued to erode and weather although its interval of most intense weathering had passed.

A different idea about southern African climatic conditions during the earlier Cenozoic was suggested by de Wit (1999, p. 736) on the basis of his work in the area around 30°S, 20°E: "more or less at the end of the Cretaceous or early [in the] Tertiary… climate changed to semi-arid conditions." This conclusion appears to stem from evidence of "a dramatic reduction in landscape denudation." It is suggested here that this reduction in denudation was, instead, a consequence of most of the existing relief having been eroded away in southern Africa by the end of Cretaceous times rather than a consequence of a decline in erosion rates resulting from a decrease in rainfall. A climatic change toward desert conditions in the western part of southern Africa came later, when the East Antarctic ice sheet first formed ca. 34 Ma (see below). Axelrod and Raven (1978) showed that by Oligocene and certainly Miocene times, when Africa had reached the position with respect to Earth's spin axis at which it has since remained stationary, the rainforest flora had migrated southward to formerly semiarid central Africa and spanned the east and west coasts. Rift-flank uplifts in the East African rift system introduced some diversity with montane components. The cool Benguela Current was responsible for the first appearance of sclerophyll-dominated vegetation in Namaqualand (Scholtz, 1983). This was part of an increasing and unprecedented bioclimatic diversification of the African continent, involving species differentiation and enrichment of biodiversity that were to continue during subsequent epochs.

Early Oligocene to Late Pliocene Times, 34–2.8 Ma

Two events in the early Oligocene changed the climate of Africa. Opening of the Southern Ocean breached land barriers joining both Australasia and South America (Drake passage) to Antarctica (Cande et al., 2000). This was followed by the onset of major ice-sheet development in the Antarctic continent, now entirely isolated by ocean. With these events, dated (from glacial deposits in ODP drill-holes in Prydz Bay—ODP Legs 178 and 188, Shipboard Scientific Party, 1999, 2000) at no later than ca. 34 Ma, an "icehouse" Earth was established. Hadley circulation was driven north by the Antarctic ice mass. The cool Benguela Current began to flow northward along the western coast of southern Africa, and the Namib Desert began to form. The Intertropical Convergence Zone migrated during the year from as far north as the Hoggar to as far south as the copper-belt of Zambia and Congo. From this time until 2.8 Ma, the southern part of Africa largely remained arid. The SE coast of South Africa appears to have been the only really wet area in the south of the continent. Farther north, the continent was much wetter, so that areas as far north as the Hoggar were subject to seasonal rainfall under the convergence zone. Establishment of the Indian Ocean monsoon by ca. 15 Ma (deMenocal, 1995), when the Zagros Seaway north of Arabia closed, and its intensification ca. 7–8 Ma (Prell and Niitsuma, 1989) subjected East Africa and Arabia to strong annual monsoonal cycles.

Global sea level was lowered by ~50 m in response to formation of the East Antarctic ice sheet and found expression through a regional increase in elevation of the African Surface over the whole of Afro-Arabia. That change can hardly have begun to take effect before the eruption at 31 Ma of the Afar plume, which pinned the slowly rotating African plate, in turn setting up shallow mantle convection and initiating Africa's presently active swells over rising, shallow-sourced plumes (Burke 2001, Burke et al., 2003a). Numerical simulations also suggest that rift-flank uplift in the East African rift system during the last ~8 m.y. drove climates in the more interior parts of the rift system toward even greater aridity than permitted by the seasonal monsoon system. This caused a lasting shift from tree-dominated to grassland-dominated land cover (Sepulchre et al., 2006), with debated consequences for the first stages of human evolution.

The relative lack of denudation on the Great Swell of southern Africa can be attributed to its elevated surface having persistently received little rainfall since uplift began in Oligocene times (Partridge, 1997). Uplift during the past 30 m.y. as a result of flexure in response to erosion of the Natal and Limpopo/Mpumalanga Drakensberg escarpments has had a powerful effect. It has stimulated orographic arrest of moisture advected from the Indian Ocean in areas close to the warm Mozambique and Agulhas Currents. The Indian Ocean high pressure cell is weaker on its western (i.e., African) side, enhancing atmospheric instability and the opportunity for rainfall to penetrate southern Africa from the east (Tyson and Preston-Whyte, 2000). The higher rainfall along the Great Escarpment on the eastern coast of South Africa has thus fostered greater erosion of the escarpment itself than of its hinterland to the west. Penetration of moisture from the Atlantic Ocean has been reduced by the changes in oceanic circulation after 34 Ma related to establishment of an Antarctic ice sheet, and by the birth of the cool Benguela Current. Adding to the high-pressure regime that prevails over continental southern Africa, the associated dry-climate patterns that followed have been promoted by the greater strength of the South Atlantic permanent high pressure cell on its eastern side compared to its western side. Therefore, in contrast to the Indian Ocean permanent anticyclone on the eastern side of southern Africa, this prevents instability and inhibits rainfall over the western side of the continent (Tyson and Preston-Whyte, 2000). Paradoxically, the Great Swell has therefore been less eroded than swells of North Africa such as the Hoggar in the middle of the Sahara, where desert conditions began to prevail for the first time only ca. 2.8 Ma (deMenocal, 1995).

Late Pliocene and Quaternary Times, since 2.8 Ma

The establishment of the Sahara Desert at 2.8 Ma has been defined with precision from an abrupt increase in the proportion of dust-borne sediment in deep-sea cores from both the Atlantic Ocean and the Arabian Sea (deMenocal, 1995). Northern Hemisphere glaciation became clearly established at about the same time (deMenocal, 1995). Oscillations between Northern Hemisphere glacial and interglacial conditions have played a dominant role in Africa's surface evolution since the radical changes at 2.8 Ma (Figs. 22–24). Because rocks and surfaces related to these more recent changes are the best preserved, that record is better known, particularly the record of the past ~900 k.y. At that time, North Atlantic sea temperatures became substantially colder, Northern Hemisphere glacial maximum ice volumes doubled, and sea-level oscillations came to exceed 100 m in ~100,000 yr cycles. Pediment gravels preserved on low-level slopes close to the coast of the Gulf of Guinea, in areas that now experience the annual visitations of the Intertropical Convergence Zone, indicate how far south desert conditions extended from the Sahara during late Quaternary glacial maxima (Brückner, 1955; Nichol, 1998). Stone implements preserved in some of the younger of those pediment gravels indicate that cycles of the past 100,000 yr have left a record (Burke and Durotoye, 1971).

THE AFRICAN SURFACE IN THE POST–30 MA BASIN-AND-SWELL REFERENCE FRAME

Burial of the African Surface in Africa's Active Continental Basins

In sedimentary basins at the continental margin, the identifiable 30 Ma surface is an unconformity that results from the combined effects of sea-level lowering when the Antarctic ice sheet first formed and the beginning of tectonic elevation of Africa's swells. That unconformity is not the African Surface itself but a new erosion surface that defines the terminal age of the surface. Offshore, on Africa's continental shelves, that surface is discernable in reflection seismic profiles in which it appears as a prominent unconformity of mid-Oligocene (30 Ma) age (see, e.g., Burke, 1996, Fig. 41 therein, and Burke et al., 2003a, p. 48).

On the continent, the African Surface is buried to shallow depths in the three large active intracontinental basins of Africa—the Kalahari, Congo, and Chad Basins—as well as in many smaller basins that also occupy low ground among Africa's presently rising topographic swells (Fig. 2). The nonmarine sedimentary rocks deposited in the basins since the surrounding swells began to rise are of eolian, lacustrine, fluvial, and colluvial origin. They are generally thin and have not proved easy to date. For instance, subsidence in the central Congo Basin was almost continuous from Paleozoic to Cenozoic times (Lawrence and Makazu, 1988), but the upwarping of the region situated between the present Congo and Chad Basins, which carries the African Surface on the lateritized plateaus of the Central African Republic, began ca. 30 Ma. The present Chad Basin subsequently started to sag and to receive the first sediments of the *continental terminal* (Lang et al., 1990) at the same time as the *Sable ocres* began to be deposited on top of the African Surface in the Congo Basin. In francophone regions of the Chad Basin, the sedimentary units above the African Surface that postdate the beginning of upward flexure of the surface ca. 30 Ma are referred to as *continental terminal* deposits. In the Nigerian part of the Chad Basin, the name Chad Formation has been used, and farther west in West Africa, the term *continental terminal* has been widely applied. In the Nigerian part of the Iullemeden Basin, the *continental terminal* is called the Gwandu Formation, and that name has been applied to similar units in several neighboring countries. Elsewhere local names abound. The Chad Basin, with an area approaching 10^6 km^2, contains a maximum of 1000 m of sedimentary rock younger than 30 Ma. It has subsided more than the other basins during the past 30 m.y., perhaps because it overlies Cretaceous rifts. Subsidence over those rifts has combined with downward flexure of the basin to accommodate sediment eroded from the surrounding rising swells. In the Chad Basin, thin Eocene nonmarine ferruginous oolites, typical of those known at outcrop on the African Surface in West Africa (Pias, 1970; Kogbe, 1981; Conrad and Lappartient, 1987), are unconformably overlain by "late Eocene" (perhaps <30 Ma?) to Holocene sediments.

In the Kalahari Basin, the term "Kalahari Group" is applied to sedimentary rocks overlying the African Surface. The Kalahari Formation outcrop extends over ~10^6 km^2 of the Kalahari Basin in Angola, Zambia, Zimbabwe, Botswana, Namibia, and South Africa. Kalahari Formation thickness averages <100 m and reaches its maximum of ~400 m only very locally over only ~2% of the Kalahari Formation's area of outcrop (Haddon, 2000).

Topographic Uplift of the African Surface on Africa's Swells

Volcanism or Absence Thereof: Its Relevance to the African Surface

Ashwal and Burke (1989) and Burke (1996, 2001) have shown that there is only one population of topographic swells on the African continent, even though some carry volcanoes and

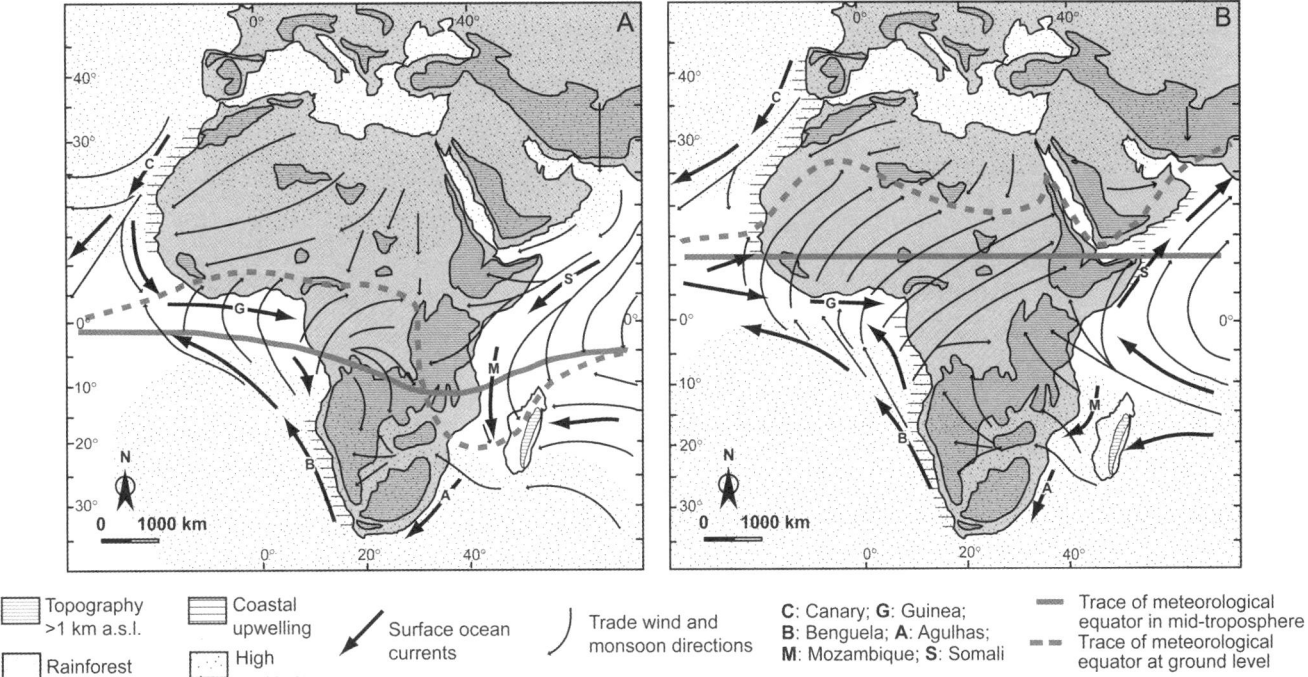

Figure 22. Seasonal climatic variations over Africa during interglacial conditions such as the present (after Leroux, 1996). (A) Mean atmospheric and moisture conditions in the lower troposphere for January and February. The Arctic and Antarctic anticyclones intermittently strengthen the subtropical high pressure cells (stipple) by detachment of cold air masses migrating equatorward. The meteorological equator rises as an inclined boundary from ground level (dashed gray line) to the mid-troposphere (solid gray line), where it becomes a vertical boundary. (B) Mean atmospheric and moisture conditions in the lower troposphere for July and August. Interglacial conditions shown in this figure for times during the past 2.8 Ma are considered to approximately match mean conditions during the 34–2.8 Ma interval.

others do not (Fig. 2). All of those swells began to rise during the past 30 m.y. The difference between swells with volcanoes and swells without volcanoes was first addressed by Kennedy (1965). He pointed out that all the continent's swells with volcanoes are on rocks that had been reactivated during Panafrican times. These were times, between ca. 700 Ma and ca. 450 Ma, when collisional fold belts accreted around the three cratonic nuclei of Africa. Volcano-free swells have developed on cratonic (>1 b.y. old) unreactivated crust. Kennedy's observation has yet to be shown wrong. His observation has two explanations: (1) The mantle lithosphere under the cratons is thick. When the base of the lithosphere is elevated under a cratonic swell, its base does not rise high enough for the pressure to be reduced to a level at which the underlying convecting mantle rock passes through the basalt solidus. (2) The sub-cratonic mantle is much more depleted than the sub-Panafrican reactivated mantle, so there is no remaining basalt in the lithospheric mantle to be melted out. Either reason would suffice to explain the absence of volcanoes, but the first reason is preferred because it relates to a more general physical process. That does not imply that the second reason is not also true.

A minority of structural swells do not form topographic elevations. Aggressive erosion close to a sea-level-controlled base level in humid climates has kept the crests of those swells close to sea level and prevented lasting surface elevation. The Senegal swell around Dakar is the best known example of this kind of structurally elevated, but topographically suppressed, swell (see Burke, 1996, Fig. 13 therein). Miocene lava flows (younger than 22 Ma) in Senegal are lateritized, implying that climatic conditions between then and the time of the first initiation of desert conditions in the Sahara (2.8 Ma) were sufficiently humid to allow laterite development. The swell of Senegal crests under Dakar where the Mammelles, the two volcanoes of Dakar, are its surface manifestation. Bellion (1987) showed that alkaline magmatism started in that area at the end of the Eocene (34 Ma) and has continued episodically until the Pleistocene, with the later lava flows in progressively less intense states of weathering. This observation has been interpreted to reflect increasing aridity during the Cenozoic and into the Pleistocene. Faulting was associated with the magmatism. A tectonic episode that has been assigned to the late Eocene (within resolution, this coincides with the ca. 30-Ma swell uplift event) was accompanied by a sudden regression of shorelines, thereby exposing a vast land surface area to weathering and erosion. The absence today of a topographic swell in Senegal, which is a low lying country, indicates that denudation by the Casamance and Senegal drainage systems was sufficient, until the Sahara Desert first formed, to keep the rising surface of the swell close to sea level. The volcanic outcrops, such as those

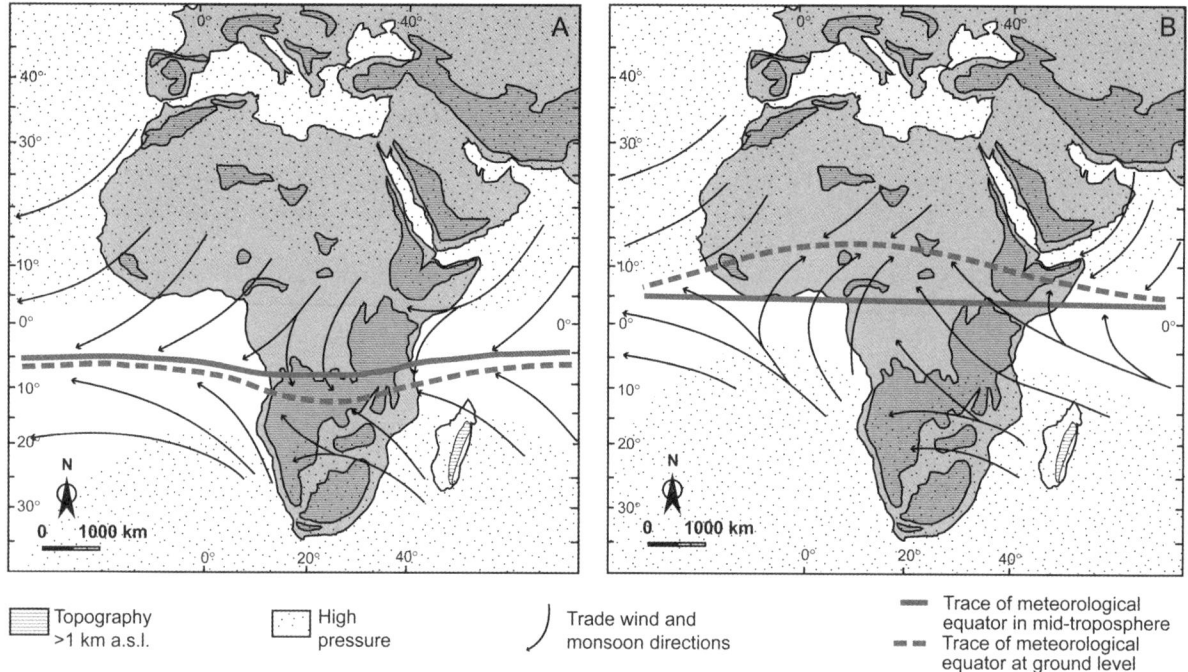

Figure 23. Atmospheric circulation over Africa during the Last Glacial Maximum for comparison with Figure 22 (modified after Leroux, 1996). (A) Mean atmospheric and moisture conditions in the lower troposphere for January and February (18–15 ka). Growth of the Arctic and Antarctic anticyclones contributed to strengthen the oceanic subtropical high pressure cells by detachment of cold air masses migrating equatorward. The meteorological equator rises as an inclined boundary from ground level (dashed gray line) to the mid-troposphere (solid gray line), where it becomes a vertical boundary. (B) Mean atmospheric and moisture conditions in the lower troposphere for July and August (18–15 ka). Under glacial conditions, the development of the Sahara Desert and an increase in rainfall in the south of the continent both in January and July are the most striking changes, compared with interglacial conditions depicted in Figure 22. Note also the more limited latitudinal band width of the tropical convergence zone compared with interglacial conditions shown in Figure 22.

of Dakar, occur on horsts (Spengler et al., 1966). The extensive, although not everywhere thick, outcrop onshore of Cenozoic *continental terminal* red beds in Senegal may represent the products of erosion of this swell during the Miocene and Pliocene. Conrad and Lappartient (1987) showed that a large proportion of the clays and sands originally considered to be *continental terminal* rocks are, in fact, deeply weathered sedimentary rocks of Miocene age that contain glauconite and marine fossils, but the bulk is terrigenous and, although deposited in a marine rather than (as previously believed) a continental environment, was sourced by the eroding swell. These rocks have been uplifted during the continuing elevation of the Dakar swell. The bulk of siliciclastic material carried to the sea by the Senegal River appears to have been deposited offshore in deep water (Jacobi and Hayes, 1982; Dombrowski et al., 2000). The lower course of the Senegal River, in its broad arc northward, gives an indication of the shape of a swell ~400 km in diameter.

The African Surface on Africa's Swells: General Characteristics

Africa's swells, on which the African Surface has been elevated episodically or continuously in the past ~30 m.y., rise typically to 1.5 ± 0.5 km a.s.l. (Burke, 1996, 2001) (Figs. 1 and 2). The 1000 × 400 km Fouta Djallon swell, as an ancient craton, carries no volcanoes but was the site of a magnitude 6.4 earthquake with a strike-slip mechanism in December 1983 (Langer et al., 1987). This is an example of a typical active African swell (Fig. 2). The preservation of the bauxite-capped African Surface in Guinea on the Fouta Djallon, in Mali, Burkina Faso, Côte d'Ivoire, Ghana, and in Nigeria on the Jos Plateau, indicates how the African Surface has been warped upward to no more than ~1.5 km on five individual swells (Burke, 1996, Fig. 29 therein). Farther east in the Cameroon line, in which African Surface elevation has been dominated by the local uplift of volcano-capped swells, maximum elevations of the African Surface are greater (Burke, 2001). Eastward, the African Surface descends toward the Central African Republic and the Congo Basin. Two swells occupy much of the Central African Republic. That on the west, which has a NE-trending axis, extends inland from near the coast in southern Cameroon, and that on the east, which has a northwesterly trend, extends from Lake Albert in the Western Rift. The two swells meet in a saddle at 19°E, which is not much more than 0.5 km a.s.l., near N'dele in the Central African Republic. Heavy rainfall under the Intertropical Convergence Zone over the past 34 m.y., and related erosion, may help to account for this low elevation. The African Surface rises again to 1.4 km on the eastern

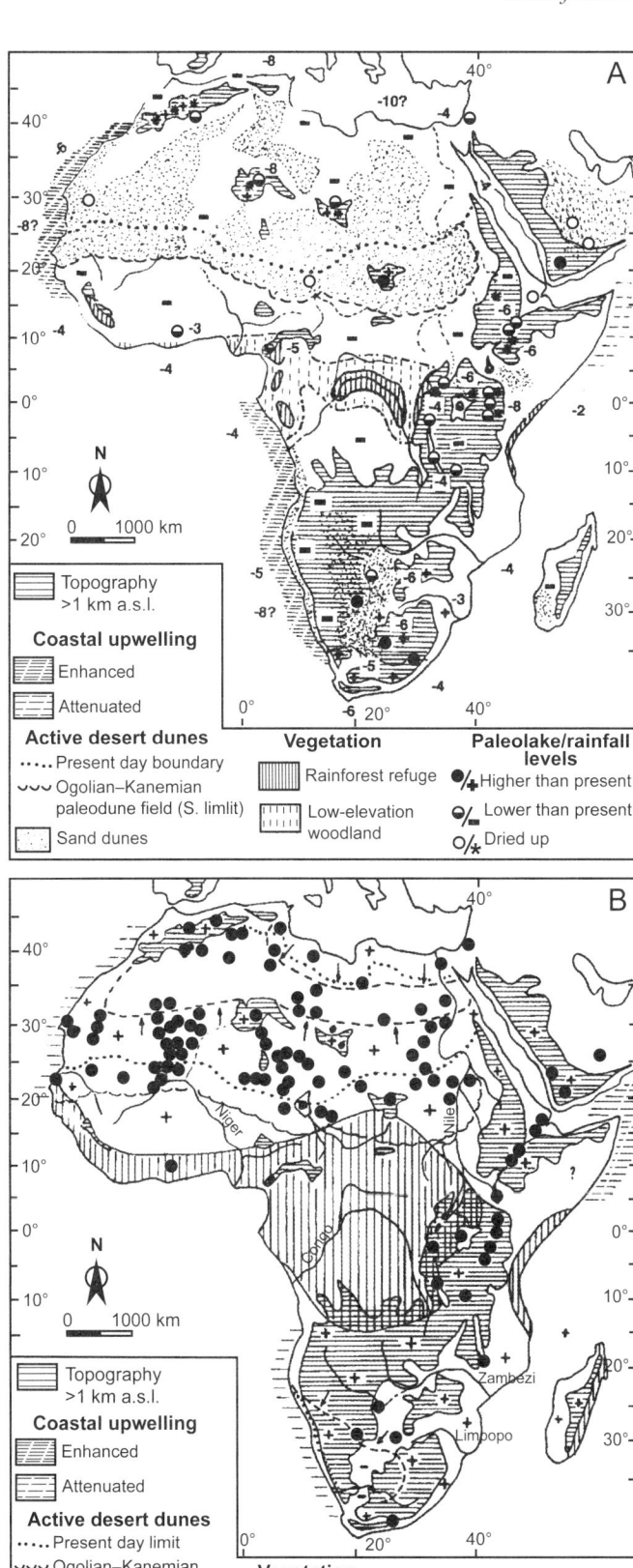

flank of the Congo Basin to become the locally named Buganda surface in Uganda at ~1.3 km a.s.l., and the Butare surface of Rwanda-Burundi at 1.7 km in an area where recent rift-shoulder uplift plays a role in the topography.

A deeply weathered (>30 m) surface is extensively preserved only on the most northern of southern African swells: the Zambian swell. The northern part of the Angolan and the Gabon swells are being intensely eroded under the climates controlled by the Intertropical Convergence Zone. Perhaps as a consequence of this, neither rises to the 1.5 km a.s.l. that appears to be typical of African swell elevation. Swells in northern Africa are generally small in area and, with the exception of those carrying major volcanoes (Hoggar, Tibesti, and Jebel Marra), do not rise very high. A possible reason may be that until the Sahara first became a desert ca. 2.8 Ma, the region was subject to heavy Intertropical Convergence Zone–related summer rainfall. In northeastern Africa, the preserved African Surface is developed across basement rocks south of Aswan. Occurrences of the African Surface at 0.9–1 km a.s.l. in the northeast are restricted to areas close to the Red Sea shoulder, which are flexed upward because they lie <200 km from hot and buoyant Red Sea active oceanic spreading centers.

More generally, the elevations of parts of several swells that reach significantly higher than the crest of our typical swell, the Fouta Djallon (1.5 km or more a.s.l.), can be attributed to processes additional to the shallow mantle convection process that is generally elevating Africa's swells (Burke, 1996, 2001; Burke et al., 2003a). In some cases the causes of additional elevation can be identified.

(1) Volcanoes and volcanic areas, as well as the African Surface beneath them, are in some places elevated above 1.5 km. This is attributable to the presence of buoyant hot rock below the volcanic area. That kind of elevation is seen most spectacularly in Kilimanjaro and in other volcanic areas on the East African swell such as Mount Kenya (Karson and Curtis, 1989). Other areas in which greater elevation is linked to volcanoes include Madagascar beneath the Tsaratanana and Ankaratra stratovolcanoes, although it is notable that the basement attains a comparable elevation at Pic Boby, which is a resistant residual landform carved out of Proterozoic syenite. The Hoggar, where the preservation of bauxite indicates greater than normal elevation, is another example. The Tibesti and Darfur in the neighborhood of the Jebel Marra volcano, and the Cameroon line of volcano-

Figure 24. Synthesis of end-member paleoenvironmental conditions in Africa since 2.8 Ma (based on Leroux, 1996). Conditions are assumed to have been roughly similar to those illustrated at each glacial/interglacial cycle. (A) Situation during the Last Glacial Maximum (18–15 ka). Negative digits indicate estimated mean paleotemperatures (in °C below mean present-day values). (B) Situation during the Holocene Climatic Optimum (ca. 0.9–0.6 ka), when paleolakes recorded all-time high water levels.

capped swells (Burke, 2001), where a differential surface uplift of 1 km can be estimated for the several individual swells, provide further examples.

(2) In many places, the shoulders of active rifts are >1 km higher than the average swell elevation of 1.5 km. This is attributable to footwall uplift, which is isostatic, and to flexure of rift shoulders above the buoyant hotter rocks that underlie the rift, even in rift sectors without volcanic activity. Elevated rift shoulders include those of the Kenya rift, the Western Rift (north of Lake Tanganyika), the Kipengere or Livingstone Mountains, and the Mulanje area of Malawi. The exceptional elevation of the Ruwenzori horst has been attributed to its position at a possible restraining bend in the Western Rift, and to its being a single giant rotated block underlain by a low-angle normal fault, but knowledge of Ruwenzori is still limited. The African Surface has been sharply upwarped to 2.5 km on the peaks of the conjugate rift shoulders of the Red Sea as a consequence of lithospheric flexure and its close proximity to the hot buoyant rocks associated with newly forming ocean floor over the past 5 m.y.

(3) At the crest of, and immediately inboard of, the Great Escarpment of southern Africa, from eastern Zimbabwe to Luanda, elevation of the highest parts of four swells (see Fig. 2) can be attributed to flexure on a length scale of 100–200 km normal to the scarp as a response to active scarp erosion. This is most prominent in areas of high rainfall.

(4) The Ethiopian swell is elevated for three reasons that work together: (*a*) Impingement on the base of the lithosphere of a deep-seated mantle plume ca. 30 Ma. This is the only such swell to be elevated on the African continent during the past ~130 m.y. (Burke, 1996; Burke et al., 2003a). The swell overlies the site at which the Afar plume generated the Ethiopian trap flood basalts during ~2 m.y. starting ca. 31 Ma. (*b*) The fluxes of heat and material from the plume. Although these declined rapidly ca. 29 Ma (Hofmann et al., 1997), the present elevation of the swell reflects the influence of whatever fluxes of heat and material persist in the "plume tail" beneath the Afar. (*c*) Active upward flexure of the shoulders of the Ethiopian, Red Sea, and Gulf of Aden rifts. The present elevation of the African Surface in the pre-basalt surface of the Ethiopian swell reaches a maximum elevation of ~2 km a.s.l., and the basalt thickness above the African Surface is ~1 km. Understandably the aggregate 3 km elevation of the Ethiopian swell greatly exceeds the canonical maximum used in this paper for swell elevations of ~1.5 km.

The Great Swell of South Africa

The way in which the Great Swell of South Africa has evolved has been lately illustrated by de Wit (1999, Fig. 20A to D therein). On the basis of de Wit's detailed work near 30°S, 20°E, there was no South African swell during Cretaceous times and neither was there a Great Escarpment. Southern Africa west of ~24°E was low lying, and drained by two major rivers: the Kalahari, or ancestral Orange River, and the Karroo River. Preserved diamond-bearing gravels of Cretaceous age, which mark the former course of the Karroo River in four places, provide evidence in support of de Wit's (1999) suggestion, as do Dingle et al.'s (1983) and Lageat's (1989b) suggestions that most of the erosion on the African Surface had been completed by Late Cretaceous time (see earlier section). Jungslager (1999, Fig. 2 therein) showed that material eroded from an extensive region of southern Africa was deposited in the ancestral Orange or Kalahari River delta, where sedimentary accumulation accompanied by growth fault development reached a maximum between ca. 90 Ma and ca. 80 Ma during Coniacian to Campanian times—straddling the Santonian tectonic episode (Guiraud and Bosworth, 1997) whose effects were so widespread throughout Africa (Burke et al., 2003a).

Less persuasive in the light of the analysis developed in this paper are de Wit's further suggestions (1) that the decline in erosion in western South Africa at the beginning of the Cenozoic marks a decline in humidity (de Wit 1999, Fig. 21 therein); and (2) that there was elevation of the southern and eastern coasts of South Africa during the early Cenozoic (de Wit 1999, Fig. 20A therein). The continent-wide evidence that we have considered shows instead the following: (1) The major change in humidity in southern Africa was associated with the establishment of the East Antarctic ice sheet ca. 34 Ma. (2) Elevation of the swells of Africa, including the Great Swell of South Africa, began ca. 30 Ma, which is not very different from the estimate made by Lageat (1989b), who placed it in the latest Eocene (ca. 36 Ma), and that of Partridge and Maud (1987), whose estimate was 20 Ma. De Wit's paleodrainage map (de Wit 1999, Fig. 20B therein) indicates that by the middle Miocene (ca. 14 Ma) the South African swell had begun to rise. De Wit's evidence seems to us compatible with the idea that the Kalahari and Karroo Rivers, which dominated a drainage pattern that had been established by Late Cretaceous times, persisting through the earlier Cenozoic, was first perturbed ca. 30 Ma when Antarctic ice-sheet formation and the beginning of swell elevation together led to sea-level lowering and to new down-cutting of the lower Orange River and of its newly formed left bank tributary, the Koa River. The Koa River captured the Karroo River, leaving the relatively short, present-day Oliphants River at the head of the Cape submarine canyon. That canyon, like most others around Africa (Burke, 1996), had first formed ca. 30 Ma when great climatic and tectonic changes were happening and sea level fell by ~50 m.

The Orange and Vaal River drainage systems have continued to flow for the past 30 m.y. on the African Surface over the gradually rising Great Swell. There has been little incision into the surface except in the southwest, in the lowest 200 km of the Orange valley below the Aughrabies falls. The extension eastward of headwaters of the Orange and Vaal drainage basins continued into Pliocene and Pleistocene times (de Wit 1999,

Figs. 20C and 20D therein), so that together they now drain ~70% of the 10^6 km^2 South African swell. That eastward extension has been a response to the continuing westward tilting of the African Surface on the swell. The tilting is itself a response to the upward isostatic flexure of the Drakensberg (Gilchrist and Summerfield, 1994), which is in turn a response to intense erosion in southern Africa's wettest region, namely the southeast coast of South Africa, in drainage basins such as that of the Tugela and other rivers that, though short, carry large seasonal sediment loads related to thunderstorm runoff during the southern summer.

Whereas the Vaal and Orange Rivers are two long rivers flowing to the west on the South African swell and involving little incision, shorter rivers, including the Tugela and tributaries of the Limpopo, flow to the eastern coast of South Africa. The Limpopo has captured rivers such as the Crocodile, which, when the South African Swell originally began to rise ca. 30 Ma, flowed to the northwest and into the Kalahari Basin. Using the ^{40}Ar/^{39}Ar method applied to a manganiferous alluvial soil in an incised valley formed after the Limpopo had captured the Kalahari-directed drainage, van Niekerk et al. (1999a, 1999b) found an age of ca. 15 Ma. The alluvium containing the dated ferromanganese nodules lies at a channelized erosional unconformity cutting the deep Waterval saprolite mantle of the African Surface. The source for the Mn ore is a ferromanganese wad interbedded with kaolinite-rich clays on the African Surface, but the argon dating provides the crystallization (and hence resetting) age of the nodules in the cut-and-fill deposits, not the inherited age of the older wad. Although work on this topic is not yet fully published (Gutzmer and Beukes, 1998), ^{40}Ar/^{39}Ar dating of Mn oxihydroxides (todorokite) gives integrated ore-forming ages of 36.3 ± 1.5 and 42.6 ± 0.7 Ma from the Smartt mine in the Kalahari manganese field of NW South Africa. Stratigraphically, the weathered horizon is overlain by the Kalahari Formation and the stratigraphic level can be correlated with the wad within the Waterval saprolite situated several hundred kilometers to the east.

If these data are integrated into a wider regional context (Beukes et al., 1999; Gutzmer and Beukes, 2000), the preliminary ages confirm that the Mn ore-forming event during the Eocene generated some of the Mn-enriched weathered mantles sealing the African Surface across the western Highveld Plateau area. Until further ages from other sites from the Kalahari manganese field are published, we suggest that the ages obtained so far probably represent the latest stages of intense chemical weathering of the African Surface. The dating of the incision of the African Surface ca. 15 Ma near Johannesburg is at a site close to the continental divide between rivers flowing west to the Atlantic Ocean and east to the Indian Ocean. That site would have responded relatively late to rejuvenation by rivers on the swell flanks. It is relatively young but consistent with the notion of surface uplift of the deeply weathered African Surface occurring earlier, but probably after 30 Ma.

The incision of the African Surface corresponds to the post–African I cycle of King (1963), and although Partridge and Maud (1987) estimated that this cycle was initiated ca. 18 Ma and Burke (1996) placed it earlier (at 30 Ma), it may be the case that Partridge and Maud (1987) and Beukes et al. (1999) have highlighted the geomorphic manifestation of the rejuvenation. On the other hand, Burke emphasized the onset of the triggering event, i.e., the beginning of swell uplift 10–12 m.y. earlier. Rather than any serious chronological discrepancy, the time difference may simply reflect the time lag in geomorphic system response to the general updoming mechanism at the few well-constrained study sites. This reflects the importance of distinguishing, as we have done in this paper, the difference between the *local age*, the *initial age*, and the *terminal age* of a geomorphic cycle, and of the land surfaces that originate from it. In a similar perspective, structural sections of the Kwanza Basin point to Miocene (instead of Oligocene) uplift and erosion of some 1–2 km (Hudec and Jackson, 2004; Lunde et al., 1992). Two-dimensional flexural restoration of the Congo and northern Angola margins has revealed Miocene uplift of the inner margin, with a maximum uplift rate occurring during Burdigalian times and a secondary event during Tortonian times (Lavier et al., 2000; Lavier et al., 2001). Thermochronology (fluid inclusions and AFT analysis) confirms the existence of Miocene uplift in the coastal basin of Gabon and Angola (Walgenwitz et al., 1990; Walgenwitz et al., 1992). Time lags in the response of sedimentary systems to swell uplift may therefore have varied from swell to swell, also depending on how well documented they are. This does not invalidate the notion of swell uplift commencing ca. 30 Ma.

Understanding the evolution of the Great Swell of South Africa depends critically on whether the Great Escarpment, which bounds much of the swell, is appreciated to be a young feature that has developed by erosion since the swell began to rise ca. 30 Ma, or whether it is considered an older feature that was essentially rejuvenated and steepened by swell-related uplift after a long period of degradation. Discussion in the next section and in Burke (1996) shows that the idea, which many authors have embraced, that the Great Escarpment dates from the time of Gondwana breakup (ca. 150 Ma) is no longer as attractive as it was when only less-complete data sets could be taken into consideration. The idea that the Great Escarpment marks a surface exhumed from beneath sedimentary rocks deposited during the Dwyka glaciation ca. 300 Ma (see, e.g., de Wit, 1999) seems to be only locally valid. The Great Escarpment is of complex origin, with different evolution patterns on the western, or drier, and eastern, or wetter, sides of the Great Swell. The existence of the Cape Fold erosional outlier—outboard of a region where the Great Escarpment extends E-W for 700 km along lat 33°S—illustrates clearly the importance of lithology and geological structure in shaping the scenery of South Africa.

In eastern South Africa, the Limpopo/Mpumalanga Drakensberg segment of the Great Escarpment varies from bold, uniform slope capped by Karroo sandstones, Proterozoic quartzites, and dolomites for a length of 240 km north and south of Sabie (25°S), to an erosional scarp without a prominent capping lithounit in Archean basement rocks, e.g., at Mbabane (26°15′S). In Natal, the Stormberg lava volcanic pile of Lesotho, which caps the Upper Triassic Cave Sandstone Formation, again confers a bold uniform slope to the Drakensberg. The escarpment loses both its elevation and boldness where, through a transitional segment between Natal and the former Transvaal, the lavas and the Cave sandstone no longer cap the basement. Overall, the diversity of the Great Escarpment is considerable. That is no better illustrated than in the observation that the Great Escarpment is absent where the larger rivers of South Africa, for example the Orange on the west coast and the Limpopo on the east coast, descend from the interior.

IMPLICATIONS OF THE CENOZOIC SWELL MODEL FOR THE HISTORY OF THE GREAT ESCARPMENTS AT AFRICA'S CONTINENTAL MARGINS

Long-Lived Escarpments Close to Rifted Continental Margins? A Return to the Historical Type Area of the African Surface in Southern Africa

Despite evidence of regional complexity, the recognition that over the past 150 m.y. Africa has experienced two major tectonic episodes—an episode of rifted margin formation and, ~100 m.y. later, an episode of basin-and-swell formation bracketing an interval of tectonic quiescence—has nevertheless yielded a hidden dividend. It provides a perspective from which to ask questions that are critical for understanding the evolution of the erosion surfaces of Africa: for how long does topography at the passive margin of an ocean remain rift-related?—and have some >1-km-high escarpments close to rifted continental margins persisted for up to ~150 m.y.?

To answer those questions, geomorphologists must wrestle with boundary and initial conditions, most of which are usually assumed but which greatly influence the outcome of models. Since the early 1990s, the latter question has been answered in the affirmative for escarpments in many parts of the world (see, e.g., references in Matmon et al., 2002), but evidence from Africa that escarpments close to rifted continental margins have formed during the post–30 Ma basin-and-swell forming episode requires that answers to this question be handled with caution. The alternative position, according to which Great Escarpments at the mature passive margins of the African continent are younger than ca. 30 Ma, might be considered provocative, and therefore is elaborated upon. We will argue that critical information about escarpment survival close to the rifted margins of continents has to come from the record of sedimentary rock deposition in the deep water off those continental margins. Next we will point out that the interpretation of the Great Escarpment of southern Africa as long lived has come from authors who have considered only a single episode of tectonic activity when the continental margin formed. Furthermore, only restricted swaths of the escarpment have been studied, the most detailed being just 40 km wide in SE South Africa (Brown et al., 2002). An interpretation at continental scale is still lacking.

The issue of initial conditions is particularly acute in southern Africa, which not only is the type area of the African Surface after the work of L.C. King, but also is commonly thought of as a type area of simple passive margin geomorphic evolution. One idea that appears to have been carried over from earlier work into the Partridge and Maud (1987) study, for example, is that the Great Escarpment of southern Africa is a topographic feature dating from the time of formation of the rifted continental margin at 140 ± 20 Ma. Earlier discussion in this paper has shown that when a full range of data sets is examined, significant problems are generated by considering that the Great Escarpment is a ~140 m.y. old topographic feature. However, the structural configuration and the rifting and igneous histories of the past 200 m.y. in South Africa are complex. Significant topographic and geologic variation exists both across and along the length of the margin. That margin is sheared on the east and south coasts and divergent on the west coast. The possible role of residual topography related to (1) the Cape Fold convergent margin and the linked Karroo foreland basin, (2) the rifts formed and active during the final assembly of Pangea (ca. 310–250 Ma; Burke et al., 2003a, Fig. 10 therein), and (3) initiation of ocean floor formation on the east coast of Africa at 170 ± 10 Ma north of 26°S are all difficult to assess. Karroo volcanism at 183 Ma, and volcanism related to the Tristan plume in eruptions at 133 Ma in Namibia, including the Etendeka basalts, must also have influenced topography. Finally the topographic influence of the several kimberlite eruption episodes during the Jurassic and Cretaceous has not been quantified. South Africa therefore has to be a centerpiece in any discussion.

The strongest evidence of the relative youth of escarpments close to rifted continental margins comes from the integration of observations on land with information from offshore. Much information relevant to understanding the erosion and geomorphology of the continents can be derived from the offshore sedimentary record. Here we present three lines of evidence that show, for the southern part of Africa, that the age of Great Escarpments close to the coast is not comparable to that of the oldest ocean floor at the margins, and that the escarpments must have developed more recently, notably during the past 30 m.y.

1. Coward et al. (1999, Figs. 4–10 therein, summarized in Fig. 5) published stratigraphic columns for basins at the margins of the South Atlantic Ocean. In those basins, the deposition of rift-facies rocks ended ca. 125 Ma, during Barremian times, when ocean floor began to form. Except in the southernmost basins, which were too far from the equator for major carbonate deposition to occur, rift-facies rocks and thick evaporites were succeeded by carbonate rock sequences 1–2 km thick. Those sequences represent the interval during which rift shoulders stood high at the continental margins of Africa and South America. Abundant siliciclastic sediment could not reach the continental margins and conditions resembled those of the Red Sea today. Siliciclastic sedimentary rocks representing material carried by rivers that had cut through the rift-shoulders began to form a significant proportion of depositional packages at the continental

margins of the South Atlantic by Albian times (112–100 Ma), and by 90 Ma (early Coniacian time) they had overwhelmed the carbonate environment in most basins. By Santonian times (84 Ma) drainage systems of a few large rivers, including the Orange and the Benue, were integrated to become the largest drainage systems on the African side of the Atlantic Ocean, dominating erosion of the low-lying continent.

Estimating the time by which rift shoulders had ceased to play a dominant role in deposition is harder in southern regions, where carbonate rocks were not developed. In basins on the west coast of Africa that do have carbonate rocks, Coward et al. (1999) found the change to be of Turonian to Coniacian age (ca. 90 ± 5 Ma). That time range finds an echo in a detailed study of part of the Orange Basin by Stevenson and McMillan (2004), who reported "a change in fluvial style after Cenomanian times ..." (ca. 94 Ma) and (in their Fig. 15) a general increase in widths and depths of filled valleys in early Coniacian times (ca. 90 Ma). Evidence from the offshore stratigraphic record does not support the idea of a long-lived escarpment, that is, an escarpment surviving for more than a few tens of millions of years, close to the edges of the South Atlantic Ocean. This line of evidence does not show that Africa's Great Escarpments are young, but it does show that escarpments that formed as the ocean began to open did not endure.

Maps of outcropping Cretaceous sedimentary rocks in low ground on the east coast and on the continental shelf off the west coast of South Africa compiled from hundreds of outcrop and shallow borehole samples reveal that the topographic Great Swell of South Africa has the geometric form of a huge anticline (Fig. 25). The Cretaceous sedimentary rocks have been flexed upward. They dip to the east off the east coast and to the west off the west coast. After the end of Cretaceous times rocks that had lain close to sea level were flexed upward to form the Great Swell. Older rocks below an Early Cretaceous unconformity on the west coast and a mid-Jurassic unconformity on the east coast occupy the core of the swell. The outcropping Cretaceous rocks occupy the limbs. Off the west coast marine and nonmarine claystones, siltstones, and sandstones, with ages between ca. 105 Ma and ca. 82 Ma (Albian to Santonian), occupy a belt ~50 km wide by ~700 km long that unconformably overlies Precambrian basement rocks within a few kilometers of the coast. The strike of the Cretaceous rocks is continuously parallel to the coast and the dip is to the west at ~1.5° (Stevenson and McMillan, 2004). Projection of the unconformity landward

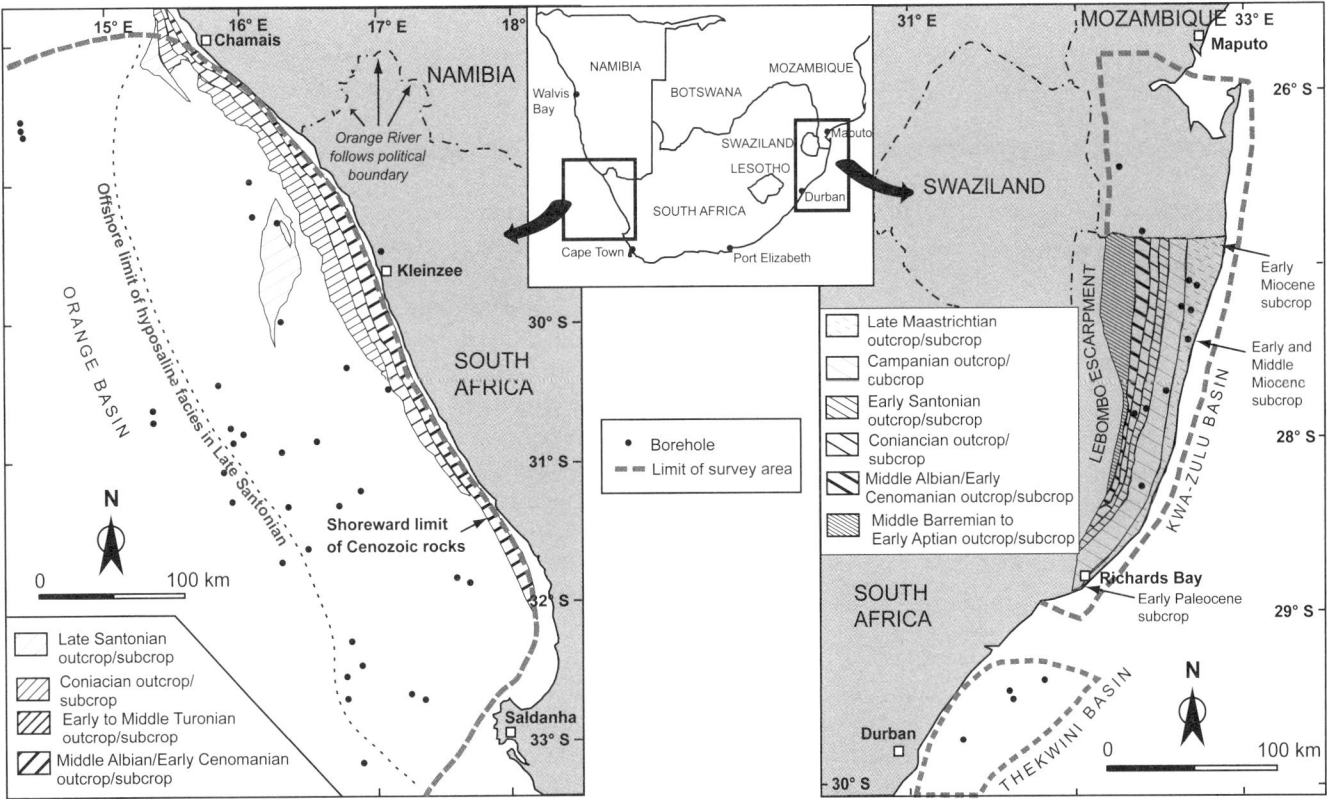

Figure 25. Southern Africa interpreted as a giant dome that stratigraphic evidence shows to have been established after the end of the Cretaceous. Cretaceous seaward-dipping sequences on both coasts, on- and offshore (adapted here from McMillan, 2003) reveal the dome, which is not as readily discerned either on its north flank, which lies on land, or off the south coast, where Cenozoic subsidence over Cretaceous rifts has complicated the record.

at constant dip takes it well above the top of the ~1-km-high Great Escarpment ~100 km inland. Horizontal river gravels containing Late Cretaceous plant fossils at Mahara Muthla on the African Surface on the top of the Great Swell, at 1.45 km a.s.l., occupy the crest of the elevated structure.

The eastern limb of the structure is known from outcrops and subcrops onshore on the east coast of South Africa between 27°S and 29°S. Cretaceous sedimentary rocks from mid-Barremian (127 Ma) to Maastrichtian (65 Ma) in age lie without significant angular unconformity on top of Lebombo mid-Jurassic lavas (184 Ma, Karroo-plume equivalent). They occupy the upper part of a structure long recognized as monoclinal (see, e.g., Davis, 1906, who wrote after observing the Drakensberg escarpment: "when the profile of the descending slope is drawn on true scale a very gentle warping without faulting seems to satisfy all the requirements of the case"); that structure is now called the Lebombo monocline. Here we provide an analysis that recognizes the Lebombo monocline as the eastern limb of a giant anticline. Lageat and Robb (1984, p. 157) point out that "the sediments of the Ecca Group [Permian] constitute a good marker as to the nature and amplitude of this monoclinal flexure. These sediments outcrop at an altitude of about 1800 m [a.s.l.] to the east of Carolina plummeting some 100 km eastward to about 300 m in the Swaziland Lowveld." The amplitude of the monoclinal flexure in the past 30 m.y. without any erosion has been 1500 m (Fig. 26). There is, therefore, no evidence of an old escarpment inherited from the time of rifting.

2. On the east coast of South Africa, McMillan's (2003) distinctive contribution has been to show that Cretaceous sedimentary rocks as young as Maastrichtian in age (ca. 65 Ma) form a concordant part of the Lebombo structure. The Cretaceous rocks involved in the structure form a belt 100 km wide (from E to W) by 200 km long (from N to S) and strike parallel to the coast. Dips in the Lebombo monocline range typically from 6° to 10° in the east. That dip projected inland would, as Davis (1906) realized, take all the Jurassic and Cretaceous rocks over the top of the ~2-km-high Great Escarpment ~100 km inland. Steeper dips and a higher Great Escarpment on the east coast are attributable to heavier rainfall, more erosion, and consequently greater upward flexure of the continental margin during the past 34 m.y. On the basis of McMillan's (2003) maps the timing of the elevation of the Great Swell of South Africa to form the core of a structure with limbs on the east and west coasts took place after deposition of the youngest rocks involved in the deformation, that is, after 65 Ma, within Cenozoic times, and hence ~60 m.y. after the oldest ocean floor was formed off both coasts. The Great Escarpment, which has cut back by erosion into the anticlinal structure on both the east and west coasts, cannot have begun to form before 65 Ma.

3. The greatest single contribution of the offshore record to recognizing that escarpments near mature passive margins are only a few tens of millions of years old comes from the evidence of recent elevation of the shore. An example from the west coast of South Africa shows Cretaceous seismic reflectors rising shoreward, which emphasizes that "Late Cretaceous sequences exhibit flexural uplift toward the coast" (Stevenson and McMillan, 2004, Fig. 2 therein). A similar illustration of upward flexure toward the coast that involves younger sedimentary rocks from the same general area is shown in Figure 27.

Evidence of the same kind from the east coast of South Africa has been obtained from a seismic profile, with well control, across the upper part of the Tugela fan (Fig. 28). Here marker horizons are picked as Oligocene (34–23 Ma) and an undated horizon is certainly Miocene in age. Both horizons rise shoreward. The upward flexure has occurred or been accentuated within the past ~30 m.y. and, as illustrated on the figure, may still be in progress. Upward flexure of sedimentary rocks toward the coast, many of which are <30 m.y. in age, can be discerned in published figures illustrating seismic lines in many places off the African coast. For example, a pulse of regional uplift of southern Africa has been detected by the solid sediment load history of the Zambezi delta, where sediment isochore maps indicate that the first Cenozoic peak of denudation since the Late Cretaceous occurred between 34 and 24 Ma (Walford et al. 2005). Other examples appear in papers by Tari et al. (2003), Karner et al. (2003), and Macgregor et al. (2003).

Modeling of the thermal histories of oil wells can yield similar results. For example, all the modeled offshore wells of southern Mozambique and those of the onshore Mozambique lowlands (except for two rift-basin related wells in Pande) have thermal histories that are compatible with a regional 1 km uplift event ca. 30 Ma (Matthews et al., 2001). We show (Fig. 29) a published interpretation of the record of the Sunray-7 well, which is one of the wells that has been modeled as showing 1 km of elevation at 30 Ma (Matthews et al., 2001). Sunray-7 lies ~150 km east of the Great Escarpment at Komatipoort. Elevation at the site

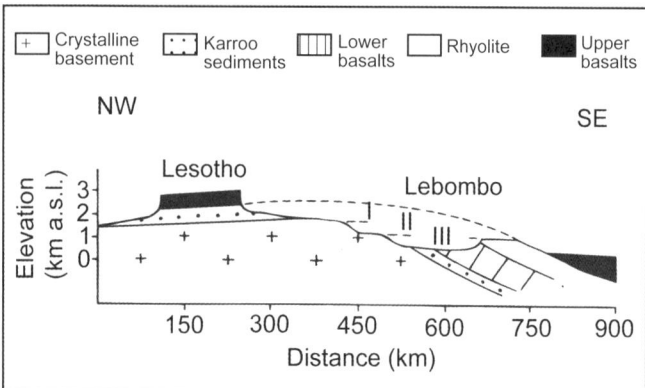

Figure 26. The Lebombo monocline near the border between Natal and Lesotho, southeastern South Africa. Successive cyclic erosion surfaces (I to III) bevel upturned geologic structures on the seaward-plunging limb of the flexure on the eastern side of the giant South African dome. The surfaces reflect progressive post–30 Ma uplift. Only II and III are strictly post-flexure cyclic surfaces. Surface I is the exhumed unconformity between the Karroo sediments and the basement. Reinterpreted after Lageat (1989b).

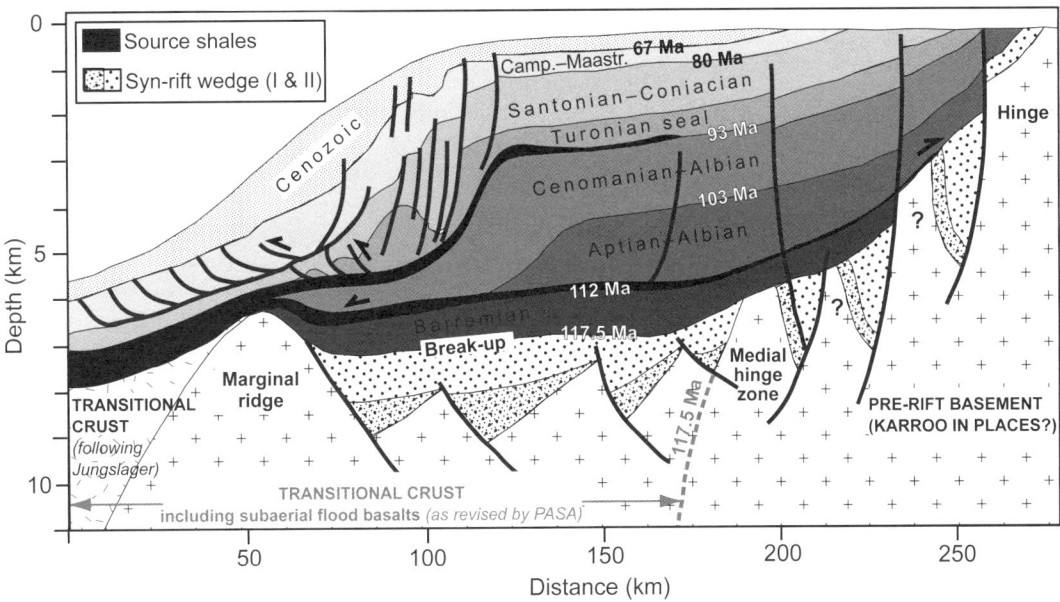

Figure 27. Schematic profile of Orange River delta sedimentary sequence, southwestern Africa (simplified after Jungslager, 1999, and Petroleum Agency of South Africa, 2003). The delta prograded onto the ocean floor between 115 and 90 Ma (continental crust of Jungslager as depicted may extend too far to the west according to the Petroleum Agency of South Africa), delivering a thick post-rift sequence of Early to Middle Cretaceous sediments (grayscale sediment packages). A major slug of sediment accumulated also during a relatively brief interval after the peak of Santonian tectonism (ca. 82 Ma), causing further progradation. Comparatively little accumulation has occurred since ca. 70 Ma because Africa was low lying between 70 and 30 Ma. After swell uplift began ca. 30 Ma, sediment flux remained limited because of the prevalence of desert conditions in South Africa west of the Drakensberg escarpment.

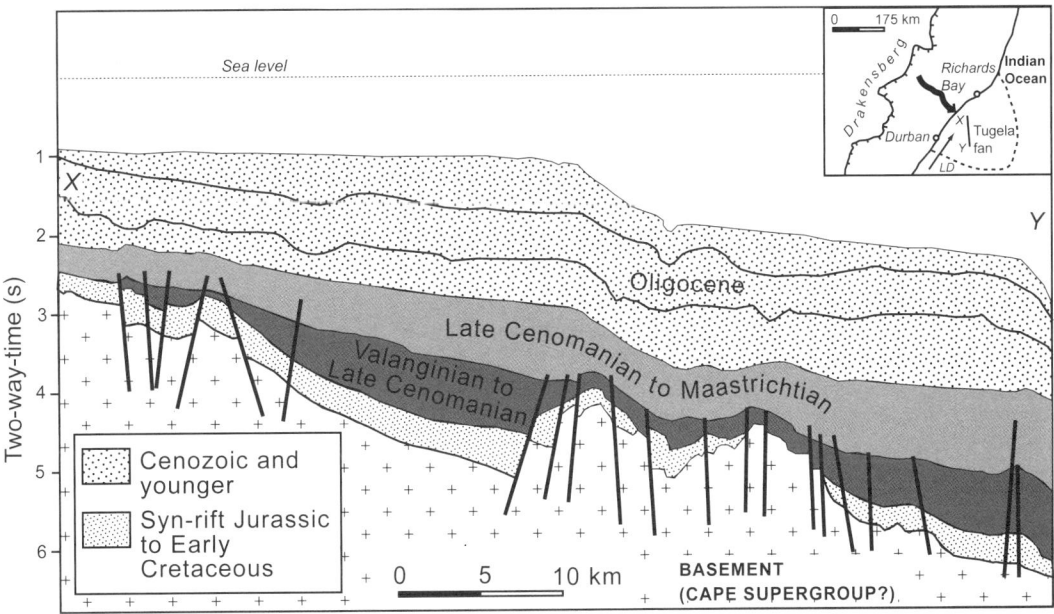

Figure 28. Cross section of the Upper Tugela Fan. Most of the sediment deposited in the fan during the past 30 m.y. has been carried north along the coast for hundreds of kilometers by longshore drift (LD) indicated by thin arrow on inset panel. Profile X-Y only extends across the shallower part of the fan, and therefore underestimates post-Eocene terrigenous input, which lies in the deep-water area extending beyond the dashed outline. Source: Petroleum Agency of South Africa (2003), modified.

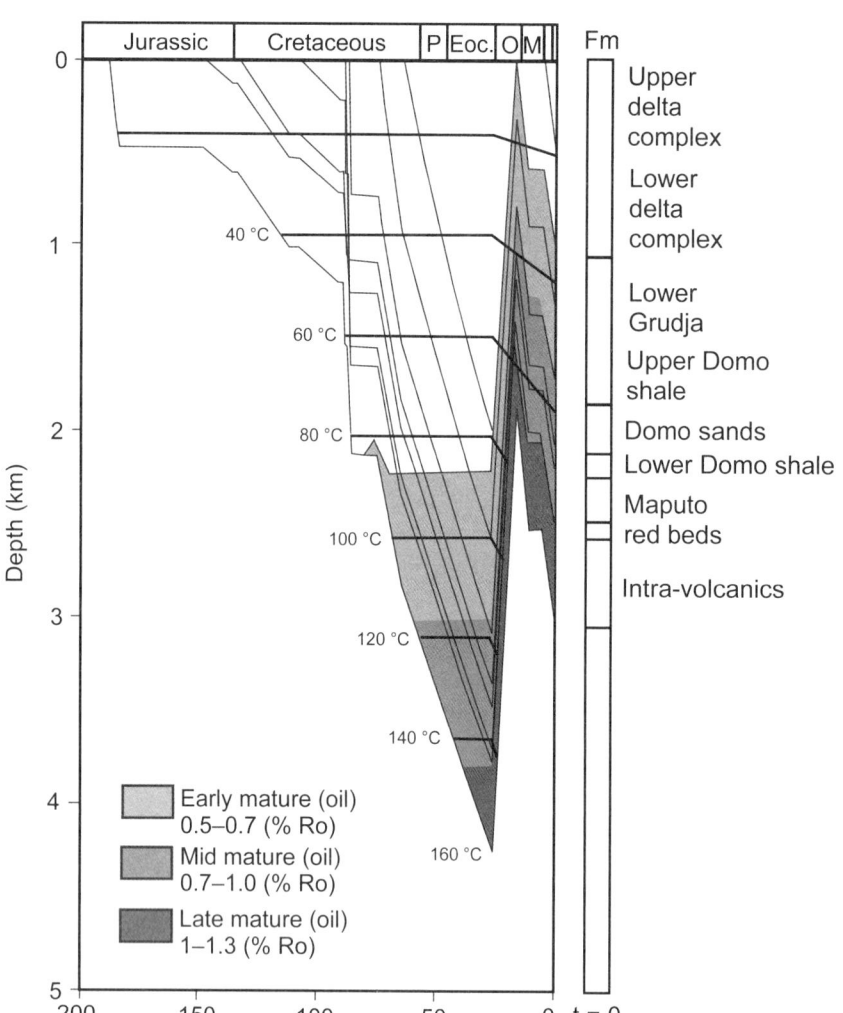

Figure 29. Thermal history of the Sunray-7 well near Xai Xai (17°S, 36°W) at the mouth of the Limpopo River (after Matthews et al., 2001). Maturity of organic material from this and other wells of coastal Mozambique required modeling involving either a regional elevation episode of ~1 km at 30 Ma or, possibly, continuing but not necessarily steady uplift starting ca. 30 Ma.

of that well is evidence that the Great Escarpment in its Lebombo sector is a young feature and a product of erosional retreat since the Great Swell of southern Africa began to rise ca. 30 Ma.

Further offshore information about the timing of elevation in areas in which escarpments exist today comes from deep-water sediment wedges off shore. The Tugela fan (Fig. 28) (Dingle et al., 1983) provides a good example. Although its load has been increased in recent time by soil erosion caused by human activity, the Tugela River today supplies 10.5×10^6 t of sediment to the coast annually. Its sediment plume, like many others on the KwaZulu-Natal coastline, extends far out into the Agulhas Current (Orme, 2005). The rise of the neighboring Drakensberg escarpment has been modeled as a response to erosion and resultant upward flexure on a ~200 km horizontal length scale (see, e.g., van der Beek et al., 2002). That kind of modeling successfully reproduces the observed geometry but does not yield information about when the flexure happened. Because evidence indicates (this study) that the elevation of swells all over Africa began ca. 30 Ma and because changes in atmospheric circulation in the Southern Hemisphere when the East Antarctic ice sheet formed led to increased rainfall on the SE coast of Africa, the erosion and the flexural uplift response of the Drakensberg are thought to have happened during the past 30 m.y. (see, e.g., Burke, 1996, Figs. 29C and D and 49C and D therein).

Offshore, few sediments have been deposited at the foot of the Natal Drakensberg escarpment because longshore drift has dispersed eroded material, and much of that transported sediment has been deposited in the Tugela fan. This interpretation is supported by the present-day nearshore dynamics of the KwaZulu coast, where the northward transport of coarser sediments is ensured by the highest wave energy of the entire 30,000 km of African coastline (>40% of wave heights exceeding 15 m), and by littoral drift in the surf zone. Finer sediment flushed farther seaward is carried south with the Agulhas Current (Orme, 1973), which has been in existence since Oligocene times. The overall result is that the Neogene sediment record on the KwaZulu shelf is not an entirely reliable source for mass-balanced calculations of Cenozoic escarpment denudation in the Drakensberg.

Dingle et al. (1983) showed that the Tugela fan does not record continuous deposition since Early Cretaceous times (ca. 125 Ma), when the ocean floor on which it lies was formed, but consists of an Early Cretaceous delta and deep-sea fan and an overlying post–30 Ma delta and deep-sea fan. Dingle's observations are compatible with the idea that the Drakensberg escarpment is a product of erosion only during the past 30 m.y. Existing seismic reflection lines (e.g., Petroleum Agency of South Africa, 2003) show that the post–30 Ma Tugela fan is a large and active depositional structure. It is associated with giant slumps into deep water. Quantification of sediment volume in deep-sea fans is almost impossible because of the extreme aspect ratios of the fans. Nevertheless, the Tugela fan data show that the fan is indeed a place in which large sediment volumes have accumulated during the past 30 m.y.

Because the Drakensberg escarpment has steep slopes, is in the right location to supply that large volume, and has experienced an appropriate rainfall regime for the past ~34 m.y., the idea that the Drakensberg formed during the past 30 m.y. seems appropriate. More work on the offshore data is needed to test this suggestion better. A prime need is for better ties between shallow water wells and seismic lines that extend from shallow water into deep water. The overwhelming majority of deep-water sedimentary basins around Africa show significant peaks of sedimentation (all greater than at any time previously) after 30 Ma (see, e.g., Burke, 1996, and D.J. Rust, in Summerfield, 1996, Fig. 1.8 therein). One exception that has been highlighted and variably interpreted in the literature (Rust and Summerfield, 1990; Lucazeau et al., 2003) could be Namibia, where Late Cretaceous and Paleogene peaks of sediment supply seem to have surpassed later ones. According to Stevenson and McMillan (2004, p. 205), Rust and Summerfield used incomplete older data from Emery and Uchupi (1984), and considered wells only on the shelf, not in deep water where much of the sediment was deposited (Fig. 30). Here we correlate the Late Cretaceous peak with material eroded during and after the Santonian tectonic event, which was recorded across all of Afro-Arabia (Guiraud and Bosworth, 1997). The post-Paleocene fall in sedimentation rates off Namibia is not attributable to the absence of post-Eocene crustal uplift. De Wit (1999) believed that aridity in SW Africa had set in as early as 65 Ma, but we argue that low erosion rates between 65 Ma and 34 Ma reflect the low lying nature of the continent rather than any direct climatic cause at that time. For instance, insignificant values of denudation (0.5–1 m × m.y.$^{-1}$) during Neogene times were attributed to intensified aridity of the Namib Desert following the establishment of the cold Benguela Current (van der Wateren and Dunai, 2001). Although most authors, including van der Wateren and Dunai, put the establishment of the Benguela Current ca. 15 Ma, this is probably a minimal age because recent ODP cruises did not drill deep enough to reach the older rocks. Older DSDP cruise results suggest an onset of cold current circulation closer to 34 Ma (Dingle et al., 1983). This is consistent with symmetric evidence in the Pacific Ocean that the cool Humboldt Current off Chile and Peru has been active since Oligocene times and responsible for

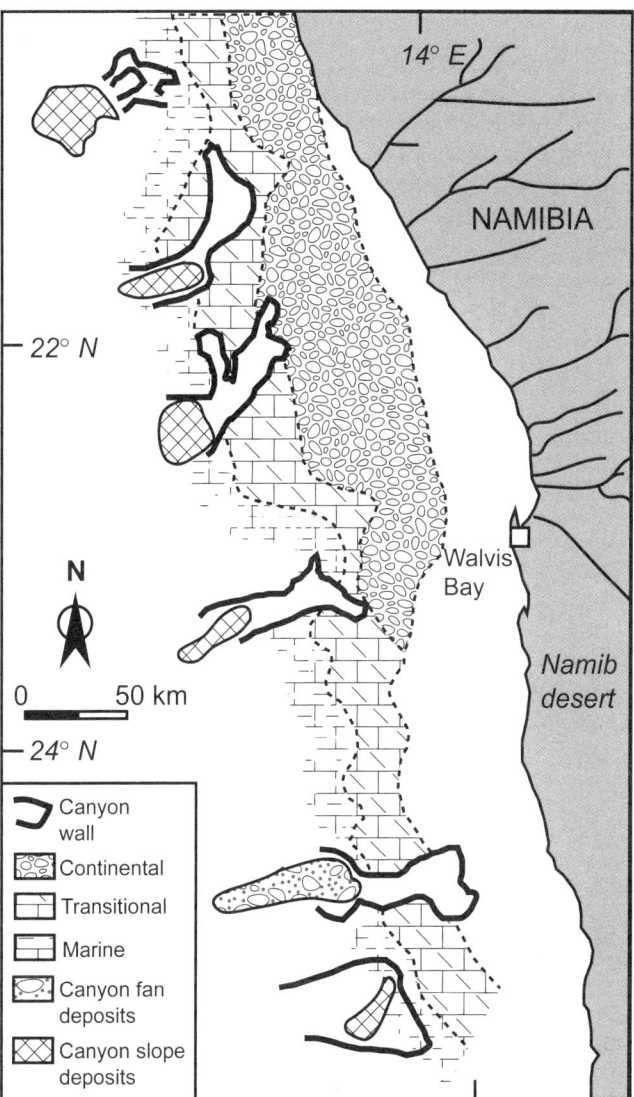

Figure 30. Post–30 Ma sediment in deep water off Namibia. Six submarine canyons off the Namib coast deposited only small (~50 km × 20 km in area) canyon base-of-slope and canyon fan deposits during the past 30 m.y. For comparison, the Congo fan, which formed over the same 30 m.y. interval offshore of a region of heavier rainfall, is 500 times larger in area (see, e.g., Anka and Séranne, 2004, and references therein). Source: Bagguley and Prosser (1999).

long-term aridity in the Andes. The result has been low erosion during the last ~30 m.y. (Dunai et al., 2005), which may explain the high topography of the Andes, where tectonic uplift in the Western Cordillera of Peru has been counterbalanced by very little denudation (Lamb and Davis, 2003).

In summary, the smaller amounts of rock eroded off South Africa to the west and south coasts during Cenozoic times can be explained by the region (1) having been low lying after the relief generated during the Santonian event was eroded, and (2) having been a semidesert for the past 34 m.y. Despite the high topographic gradient and eastward expansion of the Orange River catchment, climatic aridity starved the margin between Oligocene time and the present.

A New Perspective on the Great Escarpment of Southern Africa

Studies of long-lived topography at many passive margins have postulated either late Cenozoic uplift, in which case the topography is relatively young, or delayed topographic decay as ways of justifying its presence. The persistence of the idea that only a single tectonic event, namely rifting ca. 140 Ma, has controlled development of the Great Escarpment of Southern Africa (e.g., Gilchrist and Summerfield, 1990) is remarkable, because many who study the geomorphology of southern Africa accept that there has been a second major tectonic event, represented by uplift of the Great Swell of southern Africa, within the past 30 m.y. (see, e.g., Partridge, 1997). In some cases, evidence that is equally compatible with the two-stage tectonic event model has been analyzed only in terms of the single-event model. An example is illustrated here from Gallagher et al. (1998), who, along the lines of other authors, showed cartoon sketches of three alternative single-event tectonic models for the continental margin of southern Africa (Fig. 31).

The model in Figure 31A shows a cross section representing a "scarp retreat with erosion of rift shoulders" for the formation of an escarpment. This illustrates the point made earlier about passive margin escarpments rarely surviving more than a few tens of millions of years: the original rift shoulder disappears from the landscape. What survives is a residual topographic escarpment farther into the continent, which bears little relation to the initial rift flank scarp in magnitude, shape or geographic position. The model in Figure 31B, or "downwarp model," depicts an erosional escarpment at the edge of a flexure generated by downwarping when the passive margin formed (Jessen, 1943; Ollier and Pain, 1997). The widespread argument against this view is that the model is invalid because there was a rifting event when the Atlantic margin of Africa formed. In the downwarp model, headward erosion by rivers cuts an escarpment into a topographic ramp sloping toward the sea. We argue that this model, although not appropriate for an episode in which ocean begins to form, is appropriate for swell formation beginning at 30 Ma in the two-stage tectonic event scenario advocated here. The downwarp model of Figure 31B, which persistently lacks a horizontal scale in the literature, may not apply to fairly steep or short-wavelength flexures, but may apply to broader swell flanks of the kind described in this paper. The model in Figure 31C shows a "pinned divide" variant of the rift shoulder model A.

Two-event models were not discussed by Gallagher et al. (1998) and they appear to have been rarely considered by recent students of passive margin geomorphology. For exceptions related to Greenland and Norway, see Japsen et al. (2005) and Redfield et al. (2005). We use the sketches shown in Figure 31D to illustrate what is to be theoretically expected in a fission-track cooling pattern in a two-stage tectonic event model, first with rifting and erosion of rift shoulders, and later with formation of a swell and development of an escarpment eroded into the swell flank. Initial rift shoulder erosion is likely to have removed ~5 km of rock from the surface and, in a conductive thermal gradient of ~20 °C × km^{-1}, reset the AFT below the escarpment. Such an event is illustrated in Figure 31A. An escarpment would have

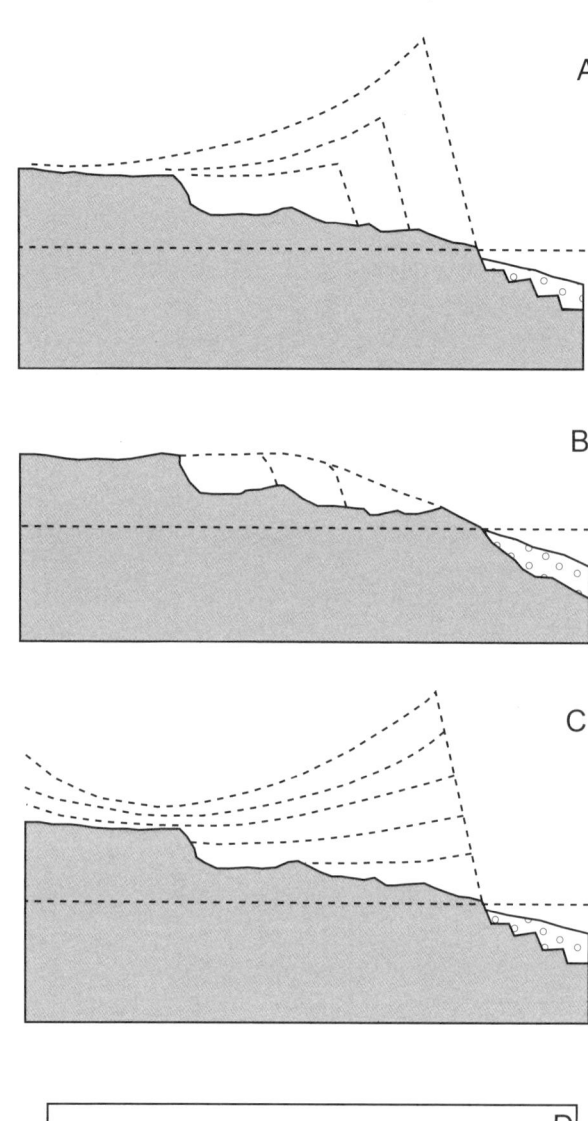

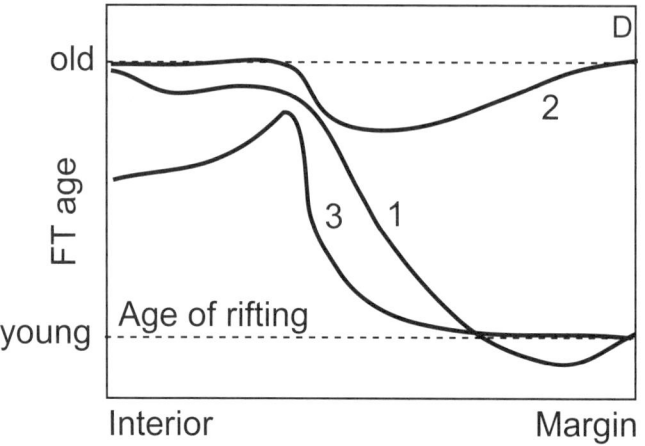

been formed, and as erosion continued inboard of the eroded rift shoulders, the retreating escarpment would have become a progressively lower feature until eventually (by ca. 70 Ma?) it became a comparatively minor topographic feature. Away from the coast, fission-track ages on the eroded surface would have remained old because no more than ~1 km would have been eroded. The AFT results along the whole profile would have been those of the curve labeled 1 in Figure 31D. The rise of the Great Swell starting at 30 Ma would have led to development of a new Great Escarpment by swell-flank erosion roughly conforming to the scenario in Figure 31B. Erosion of the swell flank by scarp retreat is likely to have removed a thickness of less than 2 km of rock and be poorly resolved by AFT. The erosion of the Great Swell during the past 30 m.y. roughly corresponds to the model of Figure 31B, although isostatic elevation onshore and isostatic depression offshore (see Burke, 1996, Fig. 26B therein) must also have been involved. Erosion of the past 30 m.y. affected the same area as the initial erosion of the rift shoulders so that the expected AFT results of the two-stage model are those of curve 1 added to curve 2 in Figure 31D. That generates a pattern very similar to that of curve 3, which is what Gallagher et al. (1998) suggested they had recognized from the AFT data. A pinned divide is perhaps a conceivable possibility if a single-event scenario must

Figure 31. Theoretical erosion and apatite fission-track (AFT) age distribution patterns at a rifted continental margin (widely reproduced in the literature after several authors, here after Gallagher et al., 1998). The three models A to C are all idealized variants of a single-stage tectonic model. Panel D is a sketch indicating that AFT results for southern Africa are less compatible with model B than they are with either A or C, and that model C fits the observations best. (A) Scarp retreat in which a rift-shoulder escarpment is maintained high isostatically during erosion as a result of the proximity of hot, buoyant, rock below the rift; later, the proximity of hot rock at the young oceanic margin generates sufficient depth of erosion to expose rocks that had zero ages before breakup. This, unfortunately, rarely happens at passive margins (Braun and van der Beek, 2004), which, on shore, most commonly expose partial annealing zones. Erosion in the interior develops as the ocean begins to widen and hot buoyant rocks no longer lie adjacent to the coast. At that time the scarp begins to retreat inland and erosion is too limited to reset AFT ages. (B) Downwarp model, in which erosion occurs nearly entirely outboard of the newly formed retreating escarpment. Given the absence of horizontal scale, this model could correspond to the erosion of a broad swell flank. (C) Pinned divide model. Given resolution limitations of the AFT method, profiles 1 and 3 in D, which illustrate cases A and C, are difficult to distinguish in practice. This may explain the equivocal nature of interpretations expressed by Gilchrist and Summerfield (1990) and later by Brown et al. (2002), who reached conclusions consistent with scarp retreat (A) but equally a pinned divide (C). Ideally, denudation in the interior is deeper in situation C than in situation A. As a consequence, a belt of low erosion at the escarpment rim can be resolved (see Fig. 32). In southern Africa, distinguishing an earlier scarp retreat event from a later swell erosion event is critical to any understanding of the topographic history. AFT cannot unequivocally discriminate a two-stage scenario in cases in which swell-flank erosion involved depths of just ~1–2 km in the past 30 m.y. without entirely resetting the AFT population.

absolutely be adhered to, but it is physically improbable because drainage headwaters recede and drainage systems reorganize by river capture over time scales of 10–100 Ma. It is also not unique or necessary to explain the data.

There is a difficulty underlying the search for a solution such as a pinned divide (Brown et al., 2002) to replace the flexural uplift and parallel scarp retreat model previously advocated for the same region (Gilchrist and Summerfield, 1990). That difficulty is to reconcile the observed sequence of stepped erosion surfaces (e.g., King, 1962; Partridge and Maud, 1987; Lageat, 1989b) with the physical understanding of isostasy. Gilchrist and Summerfield (1991, 1994) argued against the ingrained belief propagated by L.C. King, who considered that isostasy operated in discrete steps and therefore was the mechanism most appropriate to explain staircases of erosion surfaces observed seaward of continental passive margin escarpments. The argument against this (Gilchrist and Summerfield, 1991) is that isostatic processes are really continuous and therefore leave no room for episodic events in landscape development. This was argued despite conflicting claims from ten Brink and Stern (1992), who suggested that isostasy would yield different results in the case of broken rather than continuous elastic plate parameterizations, and despite widespread observations that glacio-isostatic uplift, for instance, generates discrete topographic steps (raised beach staircases) rather than long continuous ramps. Implicit in the work of Gilchrist and Summerfield was the postulate that any presumption of tiered erosion surfaces in the landscape at passive margins was an optical illusion, perhaps an ingrained article of faith largely instilled by the persuasive influence of such personalities as L.C. King.

Despite these debated issues, we suggest here that there is still room for extrinsic factors such as swell uplift. Landscape rejuvenation will permit the development of cyclic scarps between older and younger erosion surfaces at passive margins. The swell uplift mechanism in South Africa being due to a new shallow-mantle convection system, which developed ca. 30 Ma (England and Houseman, 1984; Li and Burke, 2006), implies that the cutting of younger erosion surfaces into the flanks of the post–African Surface swells, e.g., the "post-African" surfaces of King (1962) and various subsequent authors, is an achievable geomorphic outcome. The erosion surfaces mapped below the Great Escarpment are small in scale as modifiers of load distribution. Their influence on flexural isostasy of the Drakensberg Plateau edge is small and for that reason has not yet attracted analysis in physical terms. The location of some of the African plate's 75 swells younger than 30 Ma but close to sites of ca. 180–125 Ma rift-flank uplifts is a matter of observation, but its geomorphic consequences involving positive feedback in terms of uplift and fluvial response deserve to be examined in a more systematic way than has been achieved so far.

The adequate fit of the AFT results with the two-stage model advocated here can also be seen in Figure 32 (simplified from Gallagher et al., 1998), which is a map showing a summary of AFT results for southern Africa. The AFT results are not compatible with

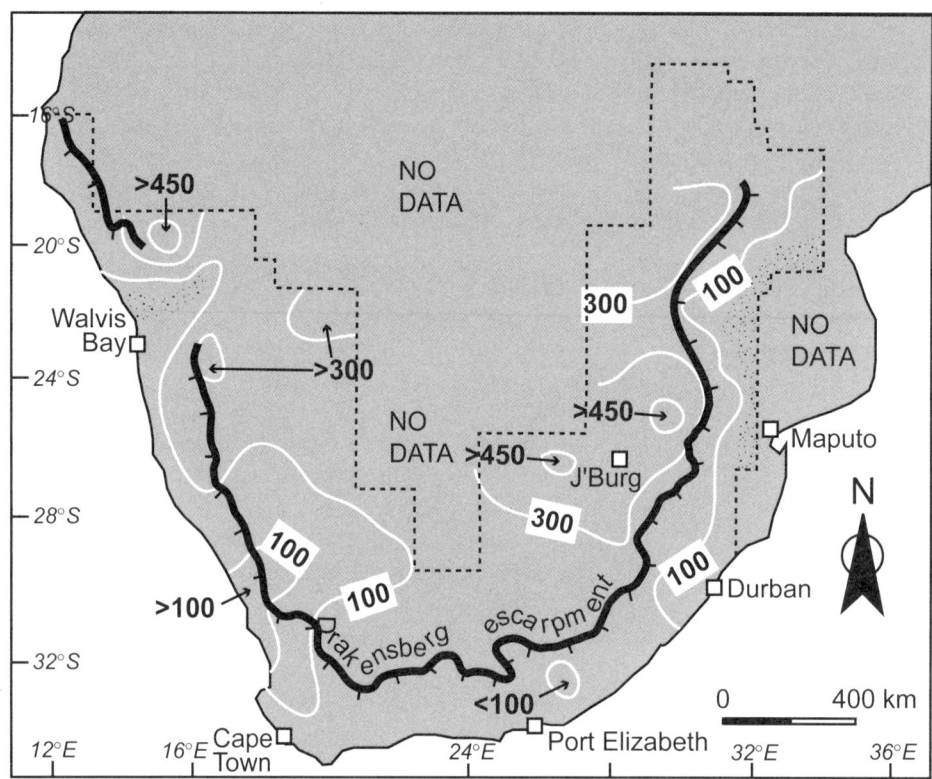

Figure 32. Map of apatite fission-track (AFT) age contours for southern Africa based on a large data set of existing point data (in Ma). Because of variable rates of cooling, the ages here are not true closure ages but need to be modeled in conjunction with mean fission-track lengths and other parameters (e.g., apatite geochemistry) to constrain the true period of rapid cooling (and therefore erosion). The older ages nevertheless define islands of low erosion above the Great Escarpment. Contours show that relatively younger ages are concentrated below the Great Escarpment, except in the west in an area where denudation in the Orange River basin appears to have rejuvenated an extensive region somewhat before 50 Ma. The few areas of most recent cooling (ca. 60 Ma; stipple) hint at the possible effect of denudation in response to uplift long after continental breakup. However, this is evident through the AFT data only in localized areas, such as coastal Namibia (see Raab et al., 2005) in an area where the current escarpment topography is subdued; and also seaward of the Transvaal Drakensberg, although few data points and possible edge effects (see the "no data" line) relating to the interpolation procedure call for more extensive research, particularly in Mozambique and Swaziland. Redrawn from Gallagher et al. (1998).

the "downwarp model" of Figure 31B. Younger ages are concentrated below the Great Escarpment, as is to be expected in the "rift shoulder erosion" and "scarp retreat" models (with or without pinning). However, this pattern is also compatible with the two-stage model preferred here, in which the coastal belt was eroded during the past 30 m.y., resulting in renewed scarp growth and retreat.

The key point is that the AFT data do not permit robust discrimination between the one- and two-stage models. The lately reported AFT borehole data (Brown et al., 2002) most unequivocally constrain a period of rapid denudation, suggesting the removal of up to 3 km of rock, between 90 and 70 Ma. We suggest that this is a reflection of the familiar Afro-Arabian continent-wide episode of tectonism and denudation during Santonian times, ca. 84 Ma (Guiraud and Bosworth, 1997) (Fig. 7), a period when many kimberlite intrusions were emplaced in South Africa. This is also in keeping with the phase of rapid exhumation between 80 and 60 Ma detected by Raab et al. (2005) in western Namibia. The sense that these were essentially denudational responses to Santonian events of short duration is corroborated by the offshore depositional record in the Kudu 9 wells drilled into the submarine "delta" of the Orange River, which at the time drained most of southern Africa. The stratigraphy indicates a depositional pulse of ~1.5 km of clastic sediments, which was deemed exceptionally thick by South African standards by McMillan (1990). The succeeding Campanian, a longer interval of time, is only 0.4 km thick.

Unequivocal detection of later denudational events in the coastal belt is difficult using AFT alone because the younger AFT ages in that belt are never younger than ca. 55 Ma (Brown et al., 2002). An illustration of this uncertainty is given by current best estimates along a narrow scarp-perpendicular swath at the foot of the Lesotho highlands, suggesting that denudation after 65 Ma ranged between 0.5 and 2 km, without any strong constraints on whether that Cenozoic denudation occurred specifically after 30 Ma or earlier (Brown et al., 2002, Fig. 12 therein). Furthermore, some samples at the coastline (i.e., 0 m a.s.l.) exhibit old ages (e.g., 114 Ma) and long mean track lengths (14.5 µm, n = 75; Brown et al., 2002). Such long lengths constrain pre-Cenozoic burial temperatures at ~40–60 °C, with these samples having already experienced several kilometers of denudation during the Cretaceous. However, due to detection limitations of the AFT method, the data do not directly record the timing of erosion that later brought such coastal rock samples from the base of the near-total fission-track stability zone (~50 °C) to the surface. Given long-term geothermal constraints from one borehole (Brown et al., 2002), magnitudes of 1–2 km of Cenozoic denudation seaward of the current escarpment are required, and we suggest this denudation peaked after 30 Ma. Similar comments could be made about other coastal-zone surface samples with >14 μm mean track lengths and ages ranging between 127 and 63 Ma.

Based on these observations, any peak of erosion occurring after 30 Ma as a result of swell uplift would require one of two possible options in terms of AFT signatures. The simpler option of deep and rapid denudation should yield (1) AFT ages of less than 30 Ma, and (2) a population of long tracks. Clearly, existing data from South Africa seaward of the Drakensberg escarpment do not show this. This means that rocks outcropping in the coastal belt today did not rise rapidly from the AFT total annealing zone (i.e., from depths below the critical 110 °C isotherm) after 30 Ma, and have not thereby experienced an equivalent depth of ~5 km of erosion during the last 30 m.y.

The more complex or indeterminate option, so called because it cannot be unequivocally demonstrated using AFT alone (see numerical sensitivity tests by Braun and van der Beek, 2004), better explains three observations: the association of 13.5–14.4 μm mean track-length values, the occurrence of some more severely annealed tracks within the distributions, and ages always greater than 55 Ma (Gallagher et al., 1998; Brown et al., 2002). This suggests that the dominant longer tracks constrain burial temperatures of <60 °C before post–30 Ma erosion drove the samples to the surface. It is well known that AFT analysis does not unequivocally detect cooling at temperatures <60 °C. Furthermore, when rapid cooling is obtained, it may be a real event in appropriate circumstances, but it may equally be an artifact generated by the annealing algorithm (Gunnell et al., 2003), or relate to apatite geochemistry or a range of operator-dependent parameters (Ketcham et al., 2007). For all these reasons, independent geologic evidence such as we provide here from offshore data or otherwise is of critical value as a test of rock cooling histories (see also, in comparable settings, Gunnell, 2003; Braun and van der Beek, 2004; Gunnell et al., 2007).

In summary, AFT analysis essentially allows broad inferences to be made about depths of denudation. Based on a number of assumptions concerning geothermal gradients, the cutting power of streams, climatic conditions, and lithospheric rigidity (see, e.g., Petit et al., 2007, and Gunnell et al., 2007), an estimate can also be made of how much rock uplift has been involved, and this estimate can be balanced against the recorded denudation. However, AFT does not directly detect magnitudes of topographic uplift, and in particular is not sensitive to Neogene and later changes in topography if depths of denudation are limited and geothermal gradients low. Furthermore, in settings where topographic uplift involves limited denudation because of climatic aridity (see earlier sections), AFT is not a tool entirely suited to drawing unequivocal conclusions about the geomorphic evolution of passive margin escarpments with the level of precision normally expected by some geomorphologists—southern Africa being no exception. For the geothermal gradients believed to be locally prevalent in South Africa today and in the past (20–25 °C [see, e.g., Brown et al., 2002]), and given the AFT ages and track length parameters observed, a situation corresponding to the more complex or indeterminate case is compatible for the area seaward of the Great Escarpment with erosion of 1–2 km since 30 Ma.

It is that erosion which has generated the Great Escarpment, even though residual topography forming lithologically or structurally controlled topography, such as the Lesotho and Lyndenburg mega-buttes, may have risen with the African Surface and generated anomalously high relief in restricted locations. This interpretation is consistent with the widely held view that lithological control is more pronounced in low-relief, weathering-limited landscapes such as those produced by long cycles of denudation in tectonically stable settings, as for example the African Surface. Clayton and Shamoon (1998), for instance, demonstrated that variations in rock resistance to erosion accounted for most of the distribution of local relief in the British Isles, which is tectonically a fairly stable area and exhibits quite low relief. By analogy, it is likely that a large proportion of local relief on African swells is lithologically controlled.

The two-stage tectonic event model proposed here—rift-flank uplift and erosion related to continental breakup, followed ~100 m.y. later by swell uplift and swell-flank erosion—singles out the "pinned divide," and, more generally, any single-event scarp evolution model, as nonunique for southern Africa. In this sense, initial rifting of Africa's passive margins bears little direct link with today's landscape. The topographic memory of the breakup of Gondwana has been lost. We consider, instead, that swell uplift after ca. 30 Ma provided gravitational potential, and hence equal opportunities for erosive mechanisms to intensify along the South African plateau edge from Mozambique to Namibia. Meanwhile, climatic asymmetry allowed the escarpment to be (1) more dynamic, that is, to grow into a bolder topographic feature, where climatic asymmetry on either side of the escarpment crest has been greatest (i.e., the southeast Drakensberg), but (2) less dynamic where aridity has affected both sides (e.g., Namibia after 15 Ma).

Faure and Lang (1991), Burke and Wells (1989), and Burke et al. (2003a, p. 50) have highlighted the drastic changes that the drainage patterns of Africa have undergone as a consequence of elevation changes in excess of 500 m that have occurred since 30 Ma when new topographic divides began to break up and redirect rivers on and near Africa's rising swells (Figs. 1 and 33). Some rivers, such as the Nile, were first formed after 30 Ma and others, for example the Niger, Congo, Oubangui, Chari, and

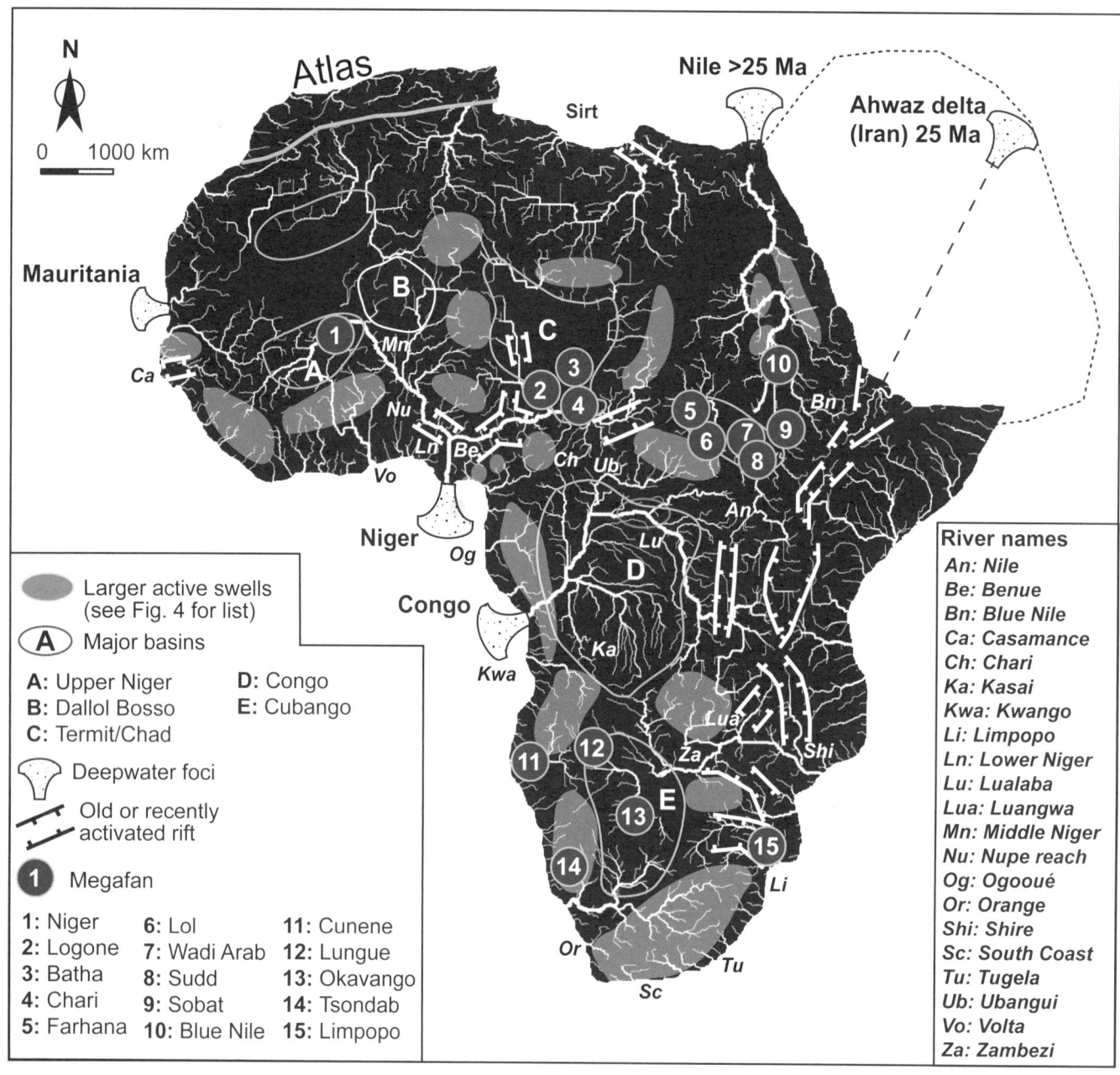

Figure 33. African rivers of the past 30 m.y. related to basins, swells, and active and reactivated rifts. Many rivers on the east and west coasts are short and flow off active swells to the Indian and Atlantic Oceans; an example is the Tugela River in southeastern South Africa. Rivers and parts of rivers flowing in old and/or reactivated rifts: Sirt (dry), Casamance, Niger (Nupe reach), Benue, Blue Nile (in part), upper Chari, Luangwa, Shire, Zambezi, Limpopo, Termit, and South Coast rift rivers. Examples of major river basins consequent on active swells are: Nile, Ahwaz (Afar swell), Congo (East African swell), upper and middle Niger flowing to inland deltas (megafan no. 1) from West African swells, Cubango (Bie Swell), Dallol Bosso (Hoggar), and Chari (Nile-Congo divide). The Orange River is the only major drainage basin with a history of more than 100 m.y. that flows on an actively rising swell. Rivers in which known drainage capture has been important during the past 30 m.y. include the Ogooué, Kasai, Kwanza, Lower Niger, Lualaba, Nupe Basin (Niger), Ubangui, Black Volta, Congo, and Zambezi. Drainage network was generated from an ETOPO2 digital elevation grid and shows Strahler stream order 5–10 (here a total of 1982 streams in growing order of line thickness). At this scale, no distinction is made between perennial and intermittent streams, and due to the nature of the digital source data and behavior of the flow grid production routines, exactness is not guaranteed in internally drained regions (e.g., deserts, rift valleys, and/or large lakes). Source for megafans: J. Wilkinson (2007, personal commun.). Topography generated from 2-minute gridded Global Relief Data (ETOPO2 v. 2, U.S. Department of Commerce, National Oceanic and Atmospheric Administration, National Geophysical Data Center, 2006, http://www.ngdc.noaa.gov/mgg/fliers/06mgg01.html).

Logone (itself under threat of capture by the Benue), experienced radical change due to river piracy and drainage diversion—the chronologies of which have been long debated (e.g., for the Niger and its knickpoint near Gao in Mali [Chudeau, 1919; Baulig, 1950; and Beaudet et al., 1977, among others]). The Zambezi delta records a phase of high sediment influx after 10 Ma, which has been attributed to a doubling of the size of the Zambezi catchment during the Pliocene (Walford et al., 2005). Longitudinal topographic profiles of African rivers exhibit the many topographic steps, defining rapids and knickzones, that are typical of rivers in the Tropics. However, in addition to these local-scale anomalies, which are often related to lithologic contrasts, the overall longitudinal profiles of large African rivers are often convex upward on wavelengths of several hundreds of kilometers that scale with the size of the post–30 Ma swells. Such convex-up profiles strongly suggest rivers that are far from equilibrium and still responding to the geologically recent or ongoing crustal deformation that this study has explored.

Detailed and systematic analysis of African drainage evolution, which involved both new river establishment and complex capture histories, still remains patchy and will require further investigation (Goudie, 2005). For instance, apart from the Okavango inland "delta," few of the megafans shown in Figure 33 are known to geologists in much detail (J. Wilkinson, 2007, personal commun.), partly because of the relative novelty of megafans as geological entities in the literature, and partly because investigation of many of these giant fans is hindered by their location in regions beset by armed conflict in recent or present times.

SUMMARY AND CONCLUSIONS

The new appreciation of the tectonic history of the African continent of the past 40 yr has made it appropriate to revisit the classic problem of the evolution of the continent's surface. The main finding has been that, to a first approximation, Africa has experienced a single, long-duration cycle of erosion since ca. 200 Ma. That cycle was initiated ca. 180 Ma following the eruption of the Karroo plume, when ocean floor began to form on the east coast of the continent and all round the Bulge of Africa. The cycle became fully established when ocean floor began to form in the South Atlantic Ocean ca. 125 Ma following the eruption of the Tristan plume. After 125 Ma, Afro-Arabia was essentially an isolated continent within the world ocean, and surrounded by Atlantic-type margins. Rift-shoulder elevations at those Atlantic-type margins, typically between 100 and 200 km wide, were eroded away in a few tens of millions of years.

Key geodynamic events that have punctuated the denudational and topographic history of Africa have been plate-pinning episodes by two mantle plumes derived from the core-mantle boundary. The first was the eruption of the Karroo plume ca. 180 Ma. The second plate-pinning event occurred ca. 30 Ma with the eruption of the Afar plume. Given the swells generated during the current (Afar-related) plume-pinning cycle, it is probable that, similarly, large topographic swells were generated at the time of the Karroo impact. Sediment supply during Jurassic and Cretaceous times came from the erosion of those swells, as well as from the erosion of short-wavelength (100–200 km) ocean-margin and intracontinental rift flanks that rose and decayed after the 145 Ma (South Atlantic) and 84 Ma (Santonian) tectonic pulses.

From time to time, sea-level and climatic changes, as well as relatively short-lived local and regional episodes of tectonism, modified the erosional processes that were active on the Afro-Arabian continental surface. As a result, the surface generated during the long cycle of erosion, which is here called the African Surface, has become a composite surface. Regional diversity of the composite African Surface reflects not only tectonic and climatic variation but also, in many areas, the role of the exhumation of erosional and depositional surfaces that had long been buried under sedimentary rocks more than 180 m.y. old.

Afro-Arabia's long cycle of erosion, which had occupied the interval during which the African Surface developed, ended abruptly ca. 30 Ma when establishment of the present basin-and-swell structure of the continent began. The effects of that tectonic event were complemented by the radical sea-level and climatic changes that had accompanied the initial establishment of the East Antarctic ice sheet some 5 m.y. earlier. Within the past 30 m.y., parts of the African Surface have been buried under the sediments of Africa's newly forming interior basins, but in many areas—except where it has been buried under a volcanic pile, such as in Ethiopia—the African Surface is well exposed on the adjacent swells.

Much of this paper is devoted to the analysis of parts of a huge regional literature that addresses the erosional history of many of those swells. That analysis focuses on the erosion surface that is here called the African Surface, although that surface goes by many different local names in many different parts of Africa. The way in which the regional literature has been treated in this paper is likely, and not without some justification, to be regarded by local experts as simplistic or even possibly cavalier. Local studies have typically identified several distinct erosion surfaces where this paper describes a single, although composite, African Surface. The justification for the "single but composite surface" approach is that it is capable of hierarchical refinement in places where adequate information exists about the composite elements of the African Surface, but it can be used to describe a single surface in the many regions where that kind of refinement is not presently possible. For reasons of space, clarity, and internal coherence to the model we propose for Africa, hierarchical refinement has not been attempted in this paper.

Laterites and bauxites occupy huge areas of the African Surface on many of Africa's actively rising swells. Many of the laterites and bauxites of Afro-Arabia, particularly the thickest examples, developed on the African Surface during the great global bauxite-forming episode between 70 Ma and 40 Ma; but not all bauxites formed during that interval. Petrological and mineralogical properties of individual bauxite occurrences allow those formed at other times to be distinguished. A large population of laterites has formed during the 30 m.y. since Africa's swells began to rise. Distinctive properties of members of that population include occurrence on the flanks rather than on the crests of swells, and the inclusion of detrital fragments of older laterites and bauxites. In many areas, the distinction of this population of rocks has not yet been attempted. If that can be done widely, it is likely to greatly facilitate refinement of the analysis attempted here.

Climatic controls on erosion surface evolution have long been recognized to be of great importance. In this paper we have shown that tectonic controls—first the establishment of rifted continental margins, then a period of relative tectonic quiescence and finally an episode of tectonically induced basin-and-swell formation—have dominated the evolution of the African Surface. However, the importance of climatic variation for the development of the African Surface is undeniable. Our approach to analyzing the role of climatic variation for the development of the African Surface has been to attempt to characterize climatic variation through time. Four intervals, 140–65 Ma, 65–34 Ma, 34–2.8 Ma, and 2.8 Ma to the present, have been distinguished.

Analyzing the development of the African Surface in the way attempted here has benefited greatly from the assimilation of offshore marine information, particularly information about when and how much siliciclastic sediment was being deposited in deep water. It is to be expected that, in the anticipated refinement of our model, the improvement of information about the deep-water geological history of the continental margins of Africa will contribute greatly to improved understanding.

The basin-and-swell structure of Africa has been a subject of investigation and discussion for nearly a century. The development of fully 75 swells in Afro-Arabia and on the oceanic part of the African plate since ca. 30 Ma corresponds to dynamic uplift generated by a new shallow-mantle convection system. The floors of the young basins—Taoudeni, Niger, Chad, Sudd, Congo, and Kalahari—cradled among the swells, lie several hundreds of meters above sea level. An unusual objective of this paper is to interrelate long-term geomorphology, mantle processes, and earth history through the analysis of a vast continent framed by unique boundary conditions. In doing so, key indicators for the analysis of other fragments of Pangea, which have different histories, different sizes, and different boundary conditions, may be discerned.

Modern tectonic understanding makes it impossible to follow the global approach of geomorphologists such as L.C. King who, making Africa the type area for a model of geomorphology and epeirogeny, assumed that other continents had followed identical patterns of landscape development. One conclusion from our study is therefore that either global geomorphic schemes such as proposed by King (1962), according to which staircases of erosion surfaces can be correlated among continents, are unlikely (because the kinematics of each plate are different, and the effects of driving mechanisms such as mantle plumes are different through space and time), or else the similarities in morphology are fortuitous and require different sets of causative mechanisms.

Epeirogeny, that is, the causes and timing of slow, persistent, and long-wavelength crustal uplifts in intraplate settings, is one of the oldest mysteries of geology (Gilbert, 1890). Epeirogeny has been considered a thorn in the side of the theory of plate tectonics (Ellenberger, 1976; Llibroutry, 1982; Le Pichon, 1984) because unambiguous and universally applicable explanations for epeirogenic uplift and downwarp have not been accommodated. The conclusion from the work reported in this paper is that there probably is no universal explanation of epeirogeny. For example, volcano-capped topographic swells in Europe, such as the Massif Central, are currently attributed to dynamic buoyancy related to an asthenospheric and thermal anomaly involving thinned mantle lithosphere (e.g., Granet et al., 1995). The topographic updoming has interfered with pre-existing tectonic structures but has occurred rapidly and recently, i.e., since the late Miocene. Explanations involving thermal anomalies and shallow-mantle convection (e.g., Li and Burke, 2006), such as suggested in this paper for the epeirogenic evolution of a single large continent over 200 m.y., may therefore be as much as can be hoped for in the current state of the art.

Throughout this broad synthesis we have emphasized that the elevation of Africa's swells took place during the past 30 m.y. Generally we have been unable to estimate whether that elevation took place rapidly soon after 30 Ma, at an even pace during the whole 30 m.y. time, or at an accelerating pace. We have considered that swells from which rivers drain to the ocean or to an interior basin are likely to be rising still, and we have considered that swells capped by active volcanoes are also likely to be rising. Because erosion has kept its summit at sea level, the Dakar swell (see earlier section) provides the most obvious exception to the difficulty of gauging the pace of uplift. At Dakar, the swell appears to have been rising fairly steadily for ~30 m.y.

The fact that 75% of Africa's petroleum has been generated during the past 30 m.y., and that 50% of it lies in reservoirs of Eocene to Holocene age is a consequence of the erosion of newly rising swells during the past 30 m.y. Denudation has varied both across the continent and with time during those 30 m.y. The Hoggar, for example, has become subject to less frequent visits from the Intertropical Convergence Zone during the Northern Hemisphere glacial episodes of the past 2.8 m.y., and currently has been conquered by desert conditions. That has also been true for the Red Sea. Western South Africa over the past ~34 m.y. has become a semidesert. Because of decreasing rainfall and erosion, some Neogene depocenters, particularly those related to the East and South African swells, are therefore unlikely to have accumulated a sufficient thickness of sedimentary rock to have driven the older underlying sediments through the oil window by deep burial.

The work reported in this paper gives a glimpse of how, through geologic time, lithospheric plate arrest by mantle plumes interacts with the arrest of moisture-bearing air masses by crustal uplift and global plate rearrangements. Such processes reorganize oceanic circulation and sea levels in a way that steers continental denudation rates, sediment routing patterns, and ultimately the distribution of natural resources—including aluminum, iron, and hydrocarbons—in a particular pattern across the surface and subsurface of continents. The integration of geology, tectonics, geomorphology, climate, and natural resources at the continental scale, crudely attempted here, forms the core of the kind of earth system science into which modern geologic research is beginning to be assimilated.

ACKNOWLEDGMENTS

Thorough reviews and insightful suggestions from Bill Bosworth (Apache Egypt Companies, Cairo), Nick Cameron (GB Petroleum PLC, London), Jeffrey Karson (Syracuse University, Syracuse), and Uwe Reimold (Museum für Naturkunde, Berlin) helped us improve this manuscript considerably. Nick Cameron and Duncan Macgregor are especially thanked for their role in creating earlier versions of our Figures 6, 7, and 33. Very special thanks go to Justin Wilkinson (NASA, Houston) for his tips on megafans and his meticulous improvement of our prose, and to Pat Bickford and the GSA staff for their patient editing of the manuscript at GSA Books.

REFERENCES CITED

Alexandre, J., and Alexandre-Pyre, S., 1987, La reconstitution à l'aide de cuirasses latéritiques de l'histoire géomorphologique du Haut-Shaba: Zeitschrift für Geomorphologie, Suppl.-Bd., v. 64, p. 119–131.

Aleva, G.J.J., 1994, Laterites: Concepts, geology, morphology and chemistry: International Soil Reference and Information Centre, Wageningen (Netherlands), 153 p.

Al-Subbary, A.K., Nichols, G.N., Bosence, D.W.J., and Al-Kadasi, M., 1998, Pre-rift forming, peneplanation or subsidence in the southern Red Sea? Evidence from the Medj-zir Formation (Tawilah Group) of western Yemen, in Purser, B., and Bosence, D., eds., Sedimentation and Tectonics in Rift Basins: Red Sea-Gulf of Aden Case: London, Chapman and Hall, p. 119–134.

Anka, Z., and Séranne, M., 2004, Reconnaissance study of the ancient Zaïre (Congo) deep-sea fan (Zai Ango Project): Marine Geology, v. 209, p. 223–244, doi: 10.1016/j.margeo.2004.06.007.

Ashwal, L.D., and Burke, K., 1989, African lithospheric structure, volcanism and topography: Earth and Planetary Science Letters, v. 96, p. 8–14, doi: 10.1016/0012-821X(89)90119-2.

Axelrod, D.I., and Raven, P.M., 1978, Late Cretaceous and Tertiary vegetation history of Africa, in Werger, M.J.A., ed., Biogeography and ecology of southern Africa: The Hague, Balkema, p. 77–130.

Babonneau, N., Savoye, B., Cremer, M., and Klein, B., 2002, Morphology and architecture of the present canyon and channel system of the Zaïre deep-sea fan: Marine and Petroleum Geology, v. 19, p. 445–467, doi: 10.1016/S0264-8172(02)00009-0.

Badalini, G., Redfern, J., and Carr, I.D., 2002, A synthesis of current understanding of the structural evolution of North Africa: Journal of Petroleum Geology, v. 25, p. 249–258, doi: 10.1111/j.1747-5457.2002.tb00008.x.

Bagguley, J., and Prosser, S., 1999, The interpretation of passive margin depositional processes using seismic stratigraphy: Examples from offshore Namibia: Geological Society [London] Special Publication 153, p. 321–344.

Baulig, H., 1950, Essais de Géomorphologie: Paris, Les Belles Lettres, 160 p.

Beaudet, G., Coque, R., Michel, P., and Rognon, P., 1977, Y a-t-il eu capture du Niger?: Bulletin de l'Association de Geographes Français, v. 445–446, p. 215–222.

Beauvais, A., and Roquin, C., 1996, Petrological differentiation patterns and geomorphic distribution of ferricretes in Central Africa: Geoderma, v. 73, p. 63–82, doi: 10.1016/0016-7061(96)00041-9.

Belinga, S.E., 1972, L'altération des roches basaltiques et les processus de bauxitisation dans l'Adamaoua (Cameroun) [Ph.D. thesis]: Paris, University of Paris VI, 510 p.

Bellion, Y., 1987, Géodynamique post-paléozoïque de l'Afrique de l'Ouest d'après l'étude de quelques bassins sédimentaires (Sénégal, Taoudenni, Iullemeden, Tchad) [Ph.D. thesis]: Avignon, Université d'Avignon, 296 p.

Bertrand, H., and Villeneuve, M., 1989, Témoins de l'ouverture de l'Atlantique Central au début du Jurassique: les dolérites tholéiitiques continentales de Guinée (Afrique de l'Ouest): Comptes Rendus de l'Académie des Sciences, Paris, sér. 2, v. 308, p. 93–99.

Beuf, S., Biju-Duval, B., de Charpal, O., Rognon, P., Gariel, O., and Bennacef, A., 1971, Les grès du Paléozoïque inférieur au Sahara. Sédimentation et discontinuités. Evolution structurale d'un craton. Paris, Technip (Collège Science et Techniques du Pétrole), v. 18, 465 p.

Beukes, N.J., van Niekerk, H.S., and Gutzmer, J., 1999, Post Gondwana African land surfaces and pedogenetic ferromanganese deposits on the Witwatersrand at the West Wits Gold Mine, South Africa: South African Journal of Geology, v. 102, p. 65–82.

Bishop, W.W., 1958, Miocene Mammalia from the Napak Volcanics, Karamoja, Uganda: Nature, v. 182, p. 1480–1482, doi: 10.1038/1821480a0.

Bishop, W.W., 1966, Stratigraphical geomorphology, in Dury, G.H., ed., Essays in geomorphology: London, Heinemann, p. 139–176.

Bishop, W.W., and Trendall, A.F., 1967, Erosion surfaces, tectonics and volcanic activity in Uganda: Quarterly Journal of the Geological Society [London], v. 122, p. 385–420.

Bocquier, G., and Gavaud, M., 1964, Étude pédologique du Niger oriental: Dakar, Éditions ORSTOM, 267 p.

Boeglin, J., 1990, Évolution minéralogique et géochimique des cuirasses ferrugineuses de la région de Gaoua (Burkina Faso) [Ph.D. thesis]: Strasbourg, Université Louis Pasteur, 187 p.

Bohannon, R.G., 1986, Tectonic configuration of the western Arabian continental margin, southern Red Sea: Tectonics, v. 5, p. 477–499.

Bond, G., 1978, Evidence for late Tertiary uplift of Africa relative to North America, South America, Australia and Europe: Journal of Geology, v. 86, p. 47–65.

Bond, G.C., 1979, Evidence for some uplifts of large magnitude in continental platforms: Tectonophysics, v. 61, p. 285–305, doi: 10.1016/0040-1951(79)90302-0.

Bornhardt, W., 1900, Zur Oberflächengestaltung und Geologie Deutsch Ostafrikas: Berlin, Reimer.

Boudouresque, L., Dubois, D., Lang, J., and Trichet, J., 1982, Contribution à la stratigraphie et à la paléogéographie de la bordure occidentale du bassin des Iullemenden au Crétacé supérieur et au Cénozoïque (Niger et Mali, Afrique de l'Ouest): Bulletin de la Société Géologique de France, v. 24, p. 685–695.

Boulangé, B., and Millot, G., 1988, La distribution des bauxites sur le craton ouest-africain: Sciences Géologiques Bulletin (Strasbourg), v. 41, p. 113–123.

Boulangé, B., 1984, Les formations bauxitiques et latéritiques de Côte d'Ivoire, Travaux et Documents 175: Paris, Éditions ORSTOM.

Boulangé, B., and Eschenbrenner, V., 1971, Note sur la présence de cuirasses, témoins des niveaux bauxitiques et intermédiaires, Plateau de Jos, Nigéria: Bulletin de l'ASEQUA, v. 31–32, p. 83–92.

Boulangé, B., Delvigne, J., and Eschenbrenner, V., 1973, Descriptions morphoscopiques, géochimiques et minéralogiques des faciès cuirassés des principaux niveaux géomorphologiques de Côte d'Ivoire: Paris, Cahiers de l'ORSTOM, série Géologie, v. 5, p. 59–81.

Boulet, R., Fauck, R., Kaloga, B., Leprun, J.-C., Vieillefon, J., and Riquier, J., 1971, Carte des sols de l'Afrique de l'Ouest, in Atlas International de l'Afrique de l'Ouest: Paris, IGN and ORSTOM, Plate 9.

Boulvert, Y., 1996, Etude géomorphologique de la République Centrafricaine: Notice explicative de la carte à 1:1,000,000: Paris, Éditions ORSTOM, 258 p.

Boulvert, Y., 2003, Carte Morphopédologique de la République de Guinée: Paris, IRD. Scale 1:500 000.

Braun, J., and van der Beek, P., 2004, Evolution of passive margin escarpments: What can we learn from low-temperature thermochronology?: Journal of Geophysical Research, v. 109, p. F04009, doi: 10.1029/2004JF000147

Briem, E., 1989, Die morphologische und tektonische Entwicklung des Roten Meer-Grabens: Zeitschrift für Geomorphologie, v. 33, p. 485–498.

Brink, A.H., 1974, Petroleum geology of the Gabon basin: American Association of Petroleum Geologists Bulletin, v. 58, p. 216–235.

Brognon, G.P., and Verrier, G.B., 1966, Oil and geology in the Cuanza Basin of Angola: American Association of Petroleum Geologists Bulletin, v. 50, p. 108–158.

Brown, R.W., Rust, D.J., Summerfield, M.A., Gleadow, A.J.W., and De Wit, M.C.J., 1990, An accelerated phase of denudation on the south-western margin of Africa: Evidence from apatite fission track analysis and the offshore sedimentary record: Nuclear Tracks and Radiation Measurements, v. 17, p. 339–350.

Brown, R.W., Summerfield, M.A., and Gleadow, A.J.W., 2002, Denudational history along a transect across the Drakensberg Escarpment of southern Africa derived from apatite fission track thermochronology: Journal of Geophysical Research, v. 107, no. B12, p. 2350, doi: 10.1029/2001JB000745

Brückner, W.D., 1955, The mantle rock (laterite) of the Gold Coast and its origin: Geologische Rundschau, v. 43, p. 307–327, doi: 10.1007/BF01764011.

Burke, K., 1975, Atlantic evaporites formed by evaporation of water spilled from Pacific, Tethyan and Southern oceans: Geology, v. 3, p. 613–616, doi: 10.1130/0091-7613(1975)3<613:AEFBEO>2.0.CO;2.

Burke, K., 1977, Aulacogens and continental breakup: Annual Review of Earth and Planetary Sciences, v. 5, p. 371–396, doi: 10.1146/annurev.ea.05.050177.002103.

Burke, K., 1996, The African Plate: South African Journal of Geology, v. 99, p. 341–409.

Burke, K., 2001, Origin of the Cameroon Line of volcano-capped swells: The Journal of Geology, v. 109, p. 349–362, doi: 10.1086/319977.

Burke, K., and Dewey, J.F., 1974, Two plates in Africa during the Cretaceous?: Nature, v. 249, p. 313–316, doi: 10.1038/249313a0.

Burke, K., and Durotoye, B., 1971, Geomorphology and superficial deposits related to large Quaternary climatic variations in south-western Nigeria: Zeitschrift für Geomorphologie, v. 15, p. 430–444.

Burke, K., and Torsvik, T.H., 2004, Derivation of large igneous provinces of the past 200 m.y. from long-term heterogeneities in the deep mantle: Earth and Planetary Science Letters, v. 227, p. 531–538, doi: 10.1016/j.epsl.2004.09.015.

Burke, K., and Wells, G.L., 1989, Trans-African drainage system of the Sahara: Was it the Nile?: Geology, v. 17, p. 743–747, doi: 10.1130/0091-7613(1989)017<0743:TADSOT>2.3.CO;2.

Burke, K., and Whiteman, A.J., 1972, Uplift, rifting and the breakup of Africa, in Tarling, D.H., ed., Proceedings from the NATO Conference on Continental Drift, vol. 2, Newcastle: Academic Press, London, p. 735–745.

Burke, K., and Wilson, J.T., 1972, Is the African plate stationary?: Nature, v. 239, p. 448–449.

Burke, K., Macgregor, D., and Cameron, N., 2003a, African petroleum systems: Four tectonic "aces" in the past 600 million years, in Arthur, T.J., Macgregor, D.S., and Cameron, N.R., eds., Petroleum geology of Africa: New themes and developing technologies: Geological Society [London] Special Publication 207, p. 21–60.

Burke, K., Ashwal, L.D., and Webb, S., 2003b, New way to map old sutures using deformed alkaline rocks and carbonatites: Geology, v. 31, p. 391–394, doi: 10.1130/0091-7613(2003)031<0391:NWTMOS>2.0.CO;2.

Cahen, L., 1954, La géologie du Congo belge: Liège, Éditions H. Vaillant Carmanne, 578 p.

Cahen, L., 1983, Breves précisions sur l'âge des groupes Crétacique post-Wealdien du Bassin intérieur du Congo: Tervuren, Belgium, Rapport Annuel du Departement de Géologie et de Minéralogie du Musée Royal de l'Afrique Centrale, Années 1981–1982, p. 61–72.

Cahen, L., and Lepersonne, J., 1952, Équivalence entre le système du Kalahari du Congo belge et les Kalahari beds d'Afrique australe: Mémoire de la Société Belge de Géologie, v. 4, 64 p.

Cande, S.C., Stock, J.M., Müller, R.D., and Ishihara, T., 2000, Cenozoic motion between East and West Antarctica: Nature, v. 404, p. 145–150, doi: 10.1038/35004501.

Caner, L., and Bourgeon, G., 2001, Andisols of Nilgiri highlands: New insight into their classification, age and genesis, in Gunnell, Y., and Radhakrishna, B.P., eds., Sahyadri, the great escarpment of the Indian subcontinent: Patterns of landscape development in the Western Ghats: Geological Society of India Memoir 47, p. 905–918.

Chardon, D., Chevillotte, V., Beauvais, A., Grandin, G., and Boulangé, B., 2006, Planation, bauxites and epeirogeny: One or two paleosurfaces on the West African margin?: Geomorphology, v. 82, p. 273–282, doi: 10.1016/j.geomorph.2006.05.008.

Chevallier, D., Giresse, P., Massengo, A., and Botoukou, G., 1972, Le site géologique de Brazzaville ou contribution à une notice explicative de la carte géologique de Brazzaville: Annales de l'Université de Brazzaville, v. 8C, p. 17–42.

Chorley, R.J., 1965, A re-evaluation of the geomorphic system of W.M. Davis, in Chorley, R.J., and Haggett, P., eds., Frontiers in Geographical Teaching: London, Methuen, p. 21–38.

Chudeau, R., 1919, La capture du Niger par le Taffassasset: Annales de Géographie, v. 28, no. 151, p. 52–60, doi: 10.3406/geo.1919.9372.

Claeys, E., 1947, Première étude des sables du Kalahari du Congo occidental: Bulletin de la Société Belge de Géologie, v. 56, p. 372–382.

Clayton, K., and Shamoon, N., 1998, A new approach to the relief of Great Britain, II. A classification of rocks based on relative resistance to denudation: Geomorphology, v. 25, p. 155–171, doi: 10.1016/S0169-555X(98)00038-5.

Coffin, M.F., and Rabinowitz, P.D., 1988, Evolution of the conjugate East African-Madagascan margins and the western Somali Basins: Geological Society of America Special Paper 226, 78 p.

Cogley, J.C., 1987, Hypsometry of the continents: Zeitschrift für Geomorphologie, Suppl.-Bd 53, 48 p.

Colin, F., Beauvais, A., Ruffet, G., and Hénocque, O., 2005, First $^{40}Ar/^{39}Ar$ geochronology of lateritic manganiferous pisolites: Implications for the Paleogene history of a West African landscape: Earth and Planetary Science Letters, v. 238, p. 172–188, doi: 10.1016/j.epsl.2005.06.052.

Collenette, P., and Grainger, P., 1994, Bauxite, in Mineral resources of Saudi Arabia: Jeddah, Special Publication of the Directorate of Geology and Mineral Resources, v. SP-2, p. 22–26.

Conrad, G., and Lappartient, J.-R., 1987, Le 'Continental terminal', sa place dans l'évolution géodynamique du bassin sénégalo-mauritanien durant le Cénozoïque: Journal of African Earth Sciences, v. 6, p. 45–60, doi: 10.1016/0899-5362(87)90106-0.

Cornelissen, A.K., and Verwoerd, W.J., 1975, The Bushmanland kimberlites and related rocks, in Ahrens, C.H., Dawson, J.B., Duncan, A.R., and Erlank, A.J., eds., Proceedings, First International Kimberlite Conference, Cape Town, 1975: Physics and Chemistry of the Earth: Oxford, UK, Pergamon Press, v. 9, p. 71–80.

Coward, M.P., Purdy, E.G., Ries, A.C., and Smith, D.G., 1999, The distribution of petroleum reserves in basins of the South Atlantic margins, in Cameron, N.R., Bate, R.H., and Clure, V.S., eds., The oil and gas habitats of the South Atlantic: Geological Society [London] Special Publication 153, p. 101–131.

Cox, K.G., 1980, A model for flood basalt vulcanism: Journal of Petrology, v. 21, p. 629–650.

Cox, K.G., 1993, Continental magmatic underplating, in Cox, K.G., McKenzie, D., and White, R.S., eds., Melting and melt movement within the earth: Royal Society of London, Philosophical Transactions, p. 155–166.

Dautria, J.M., and Girod, M.M., 1991, Relationships between Cainozoic magmatism and upper mantle heterogeneities and exemplified by the Hoggar volcanic area (Central Sahara, Southern Algeria), in Kampunzu, A.B., and Lubala, R.T., eds., Magmatism in extensional structural settings: Berlin, Springer, p. 250–268.

Davaille, A., Stutzmann, E., Silveira, G., Besse, J., and Courtillot, V., 2005, Convective patterns under the Indo-Atlantic 'box': Earth and Planetary Science Letters, v. 239, p. 233–252.

Daveau, S., 1960, Les plateaux du sud-ouest de la Haute-Volta: Étude géomorphologique: Travaux du Département de Géographie de la Faculté des Sciences Humaines de Dakar, v. 7, p. 1–61.

Davidson, A., 1983, The Omo River project: Reconnaissance geology and geochemistry of parts of Ilubabor, Kefa, Gemu Gofa, and Sidamo: Ethiopian Institute of Geological Surveys, Bulletin 2, 89 p.

Davidson, A., and Rex, D., 1980, Age of volcanism and rifting in southwestern Ethiopia: Nature, v. 283, p. 657–659, doi: 10.1038/283657a0.

Davis, W.M., 1906, Observations in South Africa: Geological Society of America Bulletin, v. 17, p. 377–450.

De Buyl, M., and Flores, G., 1986, The Southern Mozambique Basin: The most promising hydrocarbon province offshore East Africa, in Halbouty, M.T., ed., Future petroleum provinces of the world: Tulsa, Oklahoma, American Association of Petroleum Geologists Memoir 40, p. 399–425.

DeConto, R.M., Hay, W.W., Thompson, S.L., and Bergengren, J., 1999, Late Cretaceous climate and vegetation interactions: Cold continental interior paradox, in Barrera, E., and Johnson, C.C., eds., Evolution of the Cretaceous ocean-climate system: Geological Society of America Special Paper 332, p. 391–406.

Delvaux, D., and Wopfner, H., 1992, Some observations on the geomorphology and recent rift movements in the Livingstone Mountains, SW Tanzania: Paris, UNESCO, Geological and Economic Development Newsletter, v. 9, p. 171–173.

deMenocal, P.B., 1995, Plio-Pleistocene African climate: Science, v. 270, p. 53–59, doi: 10.1126/science.270.5233.53.

de Wit, M.C.J., 1999, Post-Gondwana drainage and the development of diamond placers in western South Africa: Economic Geology and the Bulletin of the Society of Economic Geologists, v. 94, p. 721–740.

Dingle, R.V., Siesser, W.G., and Newton, A.R., 1983, Mesozoic and Tertiary geology of southern Africa: Rotterdam, Balkema, 375 p.

Dixey, F., 1956, Erosion surfaces in Africa; some considerations of age and origin: Transactions of the Geological Society of South Africa, v. 59, p. 1–16.

Dombrowski, J., Faye, M., Bate, R.H., Cameron, N.R., and Carr, A.D., 2000, Evidence for a newly recognised petroleum system in the deep water portion of the Senegal sedimentary basin [abs.], in Petroleum systems and evolving technologies in African exploration and production: Petroleum Exploration Society of Great Britain/Geological Society [London], Abstracts of meeting, London, 16–18 May 2000, p. 9.

Doornkamp, J.C., 1968, The role of inselbergs in the geomorphology of southern Uganda: Transactions of the Institute of British Geographers, v. 44, p. 151–162, doi: 10.2307/621754.

Doornkamp, J.C., 1972, Trend-surface analysis of planation surfaces, with an East African case study, in Chorley, R.J., ed., Spatial analysis in geomorphology: London, Harper and Row, p. 247–281.

Driscoll, N.W., and Karner, G.D., 1994, Flexural deformation due to Amazon fan loading: A feedback mechanism affecting sediment delivery to margins: Geology, v. 22, p. 1015–1018, doi: 10.1130/0091-7613(1994)022<1015:FDDTAF>2.3.CO;2.

Drury, S.A., Kelley, S.P., Berhe, S.M., Collier, R.E.L.I., and Abraha, M., 1994, Structures related to Red Sea evolution in northern Eritrea: Tectonics, v. 13, p. 1371–1380, doi: 10.1029/94TC01990.

Dumont, P., 1991, Problèmes de datation des surfaces d'aplanissement au Zaïre: Bulletin de la Société Géographique de Liège, v. 27, p. 175–185.

Dunai, T.J., González Lopez, G.A., and Juez-Larré, J., 2005, Oligocene-Miocene age of aridity in the Atacama desert revealed by exposure dating of erosion-sensitive landforms: Geology, v. 33, p. 321–324, doi: 10.1130/G21184.1.

Du Toit, A., 1933, Crustal movement as a factor of the geographical evolution of South Africa: South African Geographical Journal, v. 16, p. 3–20.

Eales, H.V., Marsh, J.S., and Cox, K.G., 1984, The Karoo igneous province: An introduction, in Erlank, A.J., ed., Petrogenesis of the volcanic rocks of the Karoo Province: Geological Society of South Africa Special Publication 13, p. 1–26.

Egbogah, E.O., 1975, Height distribution of West African bauxites as an index of Neogene tectonism: Journal of the Nigerian Mining, Geological and Metallurgical Society, v. 10, p. 1–14.

Eldholm, O., and Coffin, M.F., 2000, Large igneous provinces and plate tectonics, in Richards, M.A., Gordon, R.G., and van der Hilst, R.D., eds., The history and dynamics of plate motions: Washington D.C., American Geophysical Union, p. 309–326.

Ellenberger, F., 1976, Épirogenèse et décratonisation: Bulletin du Bureau des Recherches Géologiques et Minières, section I, v. 4, p. 357–382.

Emery, K.O., and Uchupi, E., 1984, The geology of the Atlantic Ocean: New York, Springer, 1050 p.

England, P., and Houseman, G., 1984, On the geodynamic setting of kimberlite genesis: Earth and Planetary Science Letters, v. 167, p. 89–104.

Eschenbrenner, V., and Badarello, L., 1978, Etude pédologique de la région d'Odienné (Côte d'Ivoire). Carte des paysages morpho-pédologiques 74: Paris, Éditions ORSTOM.

Eschenbrenner, V., and Grandin, G., 1970, La séquence de cuirasses et ses différenciations entre Agnibilékrou (Côte d'Ivoire) et Diébougou (Burkina Faso): Paris, Cahiers de l'ORSTOM, ser. Géologie, v. 2, p. 205–245.

Eschenbrenner, V., Filleron, J.C., and Richard, J.-F., 1974, Applications en Côte d'Ivoire de l'étude de P. Michel: Annales de l'Université d'Abidjan, v. G6, p. 85–101.

FAO-UNESCO, 1976, Soil map of the world, v. VI, Africa: Paris, UNESCO, Scale, 1:5 000 000, 3 sheets.

Faure, H., 1966, Reconnaissance géologique des formations sédimentaires post-paléozoïques du Niger oriental: Orléans, Éditions du Bureau des Recherches Géologiques et Minières, Mémoire 47, 630 p.

Faure, H., and Lang, J., 1991, Dynamics of continental and paralic sedimentation in Africa—Quaternary models: Journal of African Earth Sciences, v. 12, p. 1–7, doi: 10.1016/0899-5362(91)90052-Z.

Fawcett, P.J., and Barron, E.J., 1998, The role of geography and atmospheric CO_2 in long-term climate change: Results from model simulations for the Late Permian to the present, in Crowley, T.J., and Burke, K., eds., Tectonic boundary conditions for climate reconstructions: Cambridge, UK, Cambridge University Press, p. 21–36.

Foster, D.A., and Gleadow, A.J.W., 1992, The morphotectonic evolution of rift-margin mountains in Central Kenya: Constraints from apatite fission-track thermochronology: Earth and Planetary Science Letters, v. 113, p. 157–171, doi: 10.1016/0012-821X(92)90217-J.

Foster, D.A., and Gleadow, A.J.W., 1996, Structural framework and denudation history of the flank of the Kenya and Anza Rifts: East Africa: Tectonics, v. 15, p. 258–271, doi: 10.1029/95TC02744.

Frakes, L.A., and Bolton, B.R., 1984, Origin of manganese giants: Sea-level change and anoxic-oxic conditions: Geology, v. 12, p. 83–86, doi: 10.1130/0091-7613(1984)12<83:OOMGSC>2.0.CO;2.

Frankart, R., 1983, The soils with sombric horizons in Rwanda and Burundi, in Beinroth, F.H., Neel, H., and Eswaran, H., eds., Proceedings, Fourth International Soil Classification Workshop, Kigali, Rwanda, 2–12 June 1981, Part I, Papers: Brussels, ABOS-AGCD, p. 48–64.

Fritsch, P., 1978, Chronologie relative des formations cuirassées et analyse géographique des facteurs de cuirassement au Cameroun: Travaux et Documents du Centre de Géographie Tropicale, Bordeaux, v. 33, p. 115–132.

Furon, R., 1960. Géologie de l'Afrique (2nd edition): Paris, Payot, 400 p.

Gallagher, K., Brown, R., and Johnson, C., 1998, Fission track analysis and its applications to geological problems: Annual Review of Earth and Planetary Sciences, v. 26, p. 519–572, doi: 10.1146/annurev.earth.26.1.519.

Garfunkel, Z., 1988, Relation between continental rifting and uplifting: Evidence from the Suez rift and northern Red Sea: Tectonophysics, v. 150, p. 33–49, doi: 10.1016/0040-1951(88)90294-6.

Garnero, E.J., Lay, T., and McNamara, A., 2007, Implications of lower-mantle structural heterogeneity for the existence and nature of whole-mantle plumes, in Foulger, G.R., and Jurdy, D.M., eds., Plates, plumes, and planetary processes: Geological Society of America Special Paper 430, p. 79–101, doi: 10.1130/2007.2430(05).

Garson, M.S., and Walshaw, R.D., 1969, The geology of the Mlanje area: Geological Survey of Malawi Bulletin 21.

Gavaud, M., 1966, Étude pédologique du Niger occidental: Rapport général: Dakar, Centre ORSTOM de Hann, 248 p.

Geukens, F., 1966, Geology of the Arabian peninsula: Yemen: U.S. Geological Survey Professional Paper, v. 560-B, 23 p.

Gilbert, G.K., 1890, Lake Bonneville: U.S. Geological Survey Monographs, v. 1, 438 p.

Gilchrist, A.R., and Summerfield, M.A., 1990, Differential denudation and flexural isostasy in formation of rifted-margin upwarps: Nature, v. 346, p. 739–742, doi: 10.1038/346739a0.

Gilchrist, A.R., and Summerfield, M.A., 1991, Denudation, isostasy and landscape evolution: Earth Surface Processes and Landforms, v. 16, p. 555–562, doi: 10.1002/esp.3290160607.

Gilchrist, A.R., and Summerfield, M.A., 1994, Tectonic models of passive margin evolution and their implications for theories of long-term landscape development, in Kirkby, M.J., ed., Process models and theoretical geomorphology: Chichester, UK, John Wiley and Sons, p. 55–84.

Giresse, P., 1982, La succession des sédimentations dans les bassins marins et continentaux du Congo depuis le début du Mésozoïque: Sciences Géologiques Bulletin (Strasbourg), v. 35, p. 183–206.

Giresse, P., 2005, Mesozoic–Cenozoic history of the Congo Basin: Journal of African Earth Sciences, v. 43, p. 301–315, doi: 10.1016/j.jafrearsci.2005.07.009.

Giresse, P., Jansen, F., Kouyoumontzakis, G., and Moguedet, G., 1981, Les fonds de la plateforme congolaise, le delta sous-marin du fleuve Congo: Bilan de huit ans de recherches sédimentologiques, paléontologiques, géochimiques et géophysiques: Paris, Éditions ORSTOM, Travaux et Documents 138, p. 13–45.

Godard, A., Simon-Coinçon, R., and Lagasquie, J.-J., 2001, Planation surfaces in basement terrains, in Godard, A., Lagasquie, J.-J., and Lageat, Y., eds., Basement regions [transl. by Y. Gunnell]: Heidelberg, Springer, p. 9–34.

Goudie, A., 2005, The drainage of Africa since the Cretaceous: Geomorphology, v. 67, p. 437–456, doi: 10.1016/j.geomorph.2004.11.008.

Grand, S.P., Van der Hilst, R.D., and Widyantoro, S., 1997, Global seismic tomography: A snapshot of convection in the earth: GSA Today, v. 7, p. 1–7.

Grandin, G., 1976, Aplanissements cuirasss et enrichissement des gisements de manganèse dans quelques régions de l'Afrique de l'Ouest: Paris, Éditions ORSTOM, Mémoire 82, 275 p.

Grandin, G., and Thiry, M., 1983, Les grandes surfaces continentales tertiaires des régions chaudes. Succession des types d'altération: Paris, Cahiers de l'ORSTOM, ser. Géologie, v. 13, p. 3–18.

Granet, M., Stoll, G., Dorel, J., Achauer, U., Poupinet, G., and Fuchs, K., 1995, Massif central (France): New constraints on the geodynamical evolution from teleseismic tomography: Geophysical Journal International, v. 121, p. 33–48, doi: 10.1111/j.1365-246X.1995.tb03509.x.

Greigert, J., 1966, Description des formations crétacées et tertiaires du bassin des Iullemeden (Afrique occidentale): Direction des Mines et de la Géologie, République du Niger, publication 2: Paris, Éditions BRGM, 234 p.

Guiraud, R., and Bosworth, W., 1997, Senonian basin inversion and rejuvenation of rifting in Africa and Arabia: Synthesis and application to plate-scale tectonics: Tectonophysics, v. 282, p. 39–82, doi: 10.1016/S0040-1951(97)00212-6.

Guiraud, R., Binks, R.M., Fairhead, J.D., and Wilson, M., 1992, Chronology and geodynamic setting of Cretaceous-Cenozoic rifting in West and Central Africa: Tectonophysics, v. 213, p. 227–234, doi: 10.1016/0040-1951(92)90260-D.

Gunnell, Y., 2003, Radiometric ages of laterites and constraints on long-term denudation rates in West Africa: Geology, v. 31, p. 131–134, doi: 10.1130/0091-7613(2003)031<0131:RAOLAC>2.0.CO;2.

Gunnell, Y., and Louchet, A., 2000, The influence of rock hardness and divergent weathering on the interpretation of apatite fission track denudation rates: Evidence from charnockites in South India and Sri Lanka: Zeitschrift für Geomorphologie, v. 44, p. 33–57.

Gunnell, Y., Gallagher, K., Carter, A., Widdowson, M., and Hurford, A.J., 2003, Denudation history of the continental margin of western peninsular India during the Mesozoic and Cenozoic: Earth and Planetary Science Letters, v. 215, p. 187–201, doi: 10.1016/S0012-821X(03)00380-7.

Gunnell, Y., Carter, A., Petit, C., and Fournier, M., 2007, Post-rift seaward downwarping at rifted margins: New insights from southern Oman using stratigraphy to constrain apatite fission-track and (U-Th)/He dating: Geology, v. 35, p. 647–650, doi: 10.1130/G23639A.1.

Gutzmer, J., and Beukes, N.J., 1998, High grade manganese ores in the Kalahari manganese field: characterisation and dating of ore-forming events: SAMANCOR, Confidential company report, 250 p.

Gutzmer, J., and Beukes, N.J., 2000, Asbestiform manjiroite and todorokite from the Kalahari manganese field, South Africa: South African Journal of Geology, v. 103, p. 163–174, doi: 10.2113/1030163.

Haddon, I.G., 2000, Kalahari Group sediments, in Partridge, T.C., and Maud, R.R., eds., The Cenozoic of southern Africa: Oxford, UK, Oxford University Press, p. 173–181.

Haq, B.U., Hardenbol, J., and Vail, P.R., 1987, Chronology of fluctuating sea levels since the Triassic: Science, v. 235, p. 1156–1167, doi: 10.1126/science.235.4793.1156.

Harrison, C.G.A., Miskell, K.J., Brass, G.W., Saltzman, E.S., and Sloan, J.L., II, 1983, Continental hypsography: Tectonics, v. 2, p. 357–377.

Haughton, S.H., 1963, The stratigraphic history of Africa south of the Sahara: Edinburgh, Oliver and Boyd, 365 p.

Hawthorn, J.B., 1975, Model of a kimberlite pipe, in Ahrens, L.H., Dawson, J.B., Duncan, A.R., and Erlank, A.J., eds., Proceedings, First International Kimberlite Conference, Cape Town, 1975: Physics and Chemistry of the Earth: Oxford, UK, Pergamon Press, v. 9, p. 1–15.

Hofmann, C., Courtillot, V., Féraud, G., Rochette, P., Yirgu, G., Ketefo, E., and Pik, R., 1997, Timing of the Ethiopian flood basalt event and implications for plume birth and global change: Nature, v. 389, p. 838–841, doi: 10.1038/39853.

Holmes, A.J., 1944, Principles of physical geology: Edinburgh, Thomas Nelson and Sons, 532 p.

Huber, M., Sloan, L.C., and Shellito, C., 2003, Early Paleogene oceans and climate: A fully coupled modeling approach using the NCAR CCSM, in Wing, S.L., Gingerich, P.D., Schmitz, B., and Thomas, E., eds., Causes and consequences of globally warm climates in the early Paleogene: Geological Society of America Special Paper 369, p. 25–47.

Hudec, M.R., and Jackson, M.P.A., 2004, Structural segmentation, inversion, and salt tectonics on a passive margin: Evolution of the Inner Kwanza Basin, Angola: Geological Society of America Bulletin, v. 114, p. 1222–1244, doi: 10.1130/0016-7606(2002)114<1222:SSIAST>2.0.CO;2.

Jacobi, R.D., and Hayes, D.E., 1982, Bathymetry, microphysiography and reflectivity characteristics of the West African margin between Sierra Leone and Mauritania, in von Rad, U., Hinz, K., Sarnthein, M., and Seibold, E., eds., Geology of the northeastern African continental margin: Berlin, Springer, p. 182–212.

Jaeger, P., 1953, Contribution à l'étude du modelé de la dorsale guinéenne. Les monts Loma: Revue de Géomorphologie Dynamique, v. 4, p. 105–113.

James, D.E., Fouch, M.J., VanDecar, J.C., and van der Lee, S., and Kaapvaal Seismic Group, 2001, Tectospheric structure beneath southern Africa: Geophysical Research Letters, v. 28, p. 2485–2488, doi: 10.1029/2000GL012578.

Jansa, L.F., and Weidmann, J., 1982, Mesozoic-Cenozoic development of the eastern North American and northwest African continental margins—A comparison, in von Rad, U., Hinz, K., Sarnthein, M., and Seibold, E., eds., Geology of the Northwest African Continental Margin: Berlin, Springer, p. 215–269.

Janse, A.J.A., 1975, Kimberlite and related rocks from the Nama plateau of South-West Africa, in Ahrens, C.H., Dawson, J.B., Duncan, A.R., and Erlank, A.J., eds., Proceedings, First International Kimberlite Conference, Cape Town, 1975: Physics and Chemistry of the Earth: Oxford, UK, Pergamon Press, v. 9, p. 81–94.

Japsen, P., Green, P.F., and Chalmers, J.A., 2005, Separation of Palaeogene and Neogene uplift on Nuussuaq, West Greenland: Journal of the Geological Society [London], v. 162, p. 299–314, doi: 10.1144/0016-764904-038.

Jessen, O., 1943, Die Randschwellen der Kontinente: Petermanns Geographische Mitteilungen, v. 241, p. 1–205.

Jongen, P., Leclercq, J., and Soberon, M., 1970, Carte des sols et de la végétation du Congo belge et du Ruanda-Urundi: Nord-Kivu et région du Lac Edouard: Brussels, Publications de l'Institut National pour l'Étude Agronomique du Congo, 78 p.

Jungslager, E.H.A., 1999, Petroleum habitats of the Atlantic margin of South Africa, in Cameron, N., Bate, R.H., and Clure, V.S., eds., The oil and gas habitats of the South Atlantic: Geological Society [London] Special Publication 153, p. 153–168.

Karner, G.D., and Driscoll, N.W., 1999, Tectonic and stratigraphic development of the West African and eastern Brazilian margins: Insights from quantitative basin modelling, in Cameron, N.R., Bate, R.H., and Clure, V.S., eds., The oil and gas habitats of the South Atlantic: Geological Society [London] Special Publication 153, p. 11–40.

Karner, G.D., Driscoll, N.W., and Barker, D.H.N., 2003, Syn-rift regional subsidence across the West African continental margin: The role of lower plate ductile extension, in Arthur, T.J., Macgregor, D.S., and Cameron, N.R., eds., Petroleum geology of Africa: New themes and developing technologies: Geological Society [London] Special Publication 207, p. 105–130.

Karson, J.A., and Curtis, P.C., 1989, Tectonic and magmatic processes at the eastern branch of the East Africa Rift and implications for magmatically active continental rifts: Journal of African Earth Sciences, v. 8, p. 431–453, doi: 10.1016/S0899-5362(89)80037-5.

Kellog, C.E., and Davol, F.D., 1949, An explanatory study of soil groups in the Belgian Congo: Brussels, Publications de l'Institut National pour l'Étude Agronomique du Congo, sér. Scientifique, v. 46.

Kennedy, W.Q., 1965, The influence of basement structure on the evolution of the coastal (Mesozoic and tertiary) basins, in Ion, D.C., ed., Salt basins around Africa: London, Institute of Petroleum, p. 7–16.

Ketcham, R.A., Carter, A., Donelick, R.A., Barbarand, J., and Hurford, A.J., 2007, Improved modeling of fission-track annealing in apatite: The American Mineralogist, v. 92, p. 799–810, doi: 10.2138/am.2007.2281.

Kilian, C., 1931, Des principaux complexes continentaux du Sahara. Compte Rendu Sommaire des Séances de la Société Géologique de France, p. 109–110.

King, L.C., 1951, South African Scenery (2nd edition): Oliver and Boyd, London and Edinburgh, 379 p.

King, L.C., 1962, The morphology of the earth: A study and synthesis of world scenery: London, Oliver and Boyd, 699 p.

King, L.C., 1963, South African scenery (3rd edition): London, Oliver and Boyd, 308 p.

King, L.C., 1976, Planation remnants upon high lands: Zeitschrift für Geomorphologie, v. 20, p. 133–148.

Klein, C., 1990, L'Évolution géomorphologique de l'Europe hercynienne occidentale et centrale: Aspects régionaux et essai de synthèse: Paris, Éditions CNRS, 178 p.

Klein, C., 1997, Du polycyclisme à l'acyclisme en géomorphologie: Gap (France), Ophrys, 300 p.

Kogbe, C.A., 1981, Cretaceous and Tertiary of the Iullemmeden Basin in Nigeria (West Africa): Cretaceous Research, v. 2, p. 129–186, doi: 10.1016/0195-6671(81)90007-0.

Kogbe, C.A., and Dubois, D., 1980, Economic significance of the 'Continental Terminal': Geologische Rundschau, v. 69, p. 429–436, doi: 10.1007/BF02104547.

Kohn, B.P., and Eyal, M., 1981, History of uplift of the crystalline basement of Sinai and its relation to opening of the Red Sea as revealed by fission track dating of apatites: Earth and Planetary Science Letters, v. 52, p. 129–141, doi: 10.1016/0012-821X(81)90215-6.

Lageat, Y., 1989a, La notion de forme structurale dans les socles: Bulletin de l'Association de Géographes Français, v. 1, p. 3–11.

Lageat, Y., 1989b, Le relief du Bushveld: Une géomorphologie des roches basiques et ultrabasiques: Clermont-Ferrand, Publications de la Faculté des Lettres et Sciences Humaines de l'Université Blaise-Pascal, 391 p.

Lageat, Y., 1997, Geomorphology of the Bushveld Complex: La Coruña, Caderno des Laboratorio Xeoloxico de Laxe: v. 22, p. 209–227.

Lageat, Y., and Robb, L.J., 1984, The relationships between structural landforms, erosion surfaces, and the geology of the Archean granite basement in the Barberton region, Eastern Transvaal: Transactions of the Geological Society of South Africa, v. 87, p. 141–159.

Lamb, S., and Davis, P., 2003, Cenozoic climate change as a possible cause for the rise of the Andes: Nature, v. 425, p. 792–797, doi: 10.1038/nature02049.

Lepersonne, J., 1956, Les aplanissements d'érosion du nord-est du Congo belge et des régions voisines: Académie Royale des Sciences Coloniales, v. 7, p. 1–108.

Lang, J., Kogbe, C., Alidou, S., Alzouma, K.A., Bellion, G., Dubois, D., Durand, A., Guiraud, R., Houessou, A., Romann, I., De Klasz, E., Salard-Cheboldaeff, M., and Trichet, J., 1990, The Continental terminal in West Africa: Journal of African Earth Sciences, v. 10, p. 79–99, doi: 10.1016/0899-5362(90)90048-J.

Langer, C.J., Bonilla, M.G., and Bollinger, G.A., 1987, Aftershocks and surface faulting associated with the intraplate Guinea, West Africa, earthquake of 22 December 1983: Bulletin of the Seismological Society of America, v. 77, p. 1579–1601.

Lavier, L., Steckler, M., and Brigaud, F., 2000, An improved method for reconstructing the stratigraphy and bathymetry of continental margins: Application to the Cenozoic tectonic and sedimentary history of the Congo margin: American Association of Petroleum Geologists Bulletin, v. 84, p. 923–939.

Lavier, L., Steckler, M.S., and Brigaud, F., 2001, Climatic and tectonic control on the Cenozoic evolution of the West African margin: Marine Geology, v. 178, p. 63–80, doi: 10.1016/S0025-3227(01)00175-X.

Lawrence, S.R., and Makazu, M.M., 1988, Zaïre's central basin: Prospectivity outlook: Oil and Gas Journal, v. 86, p. 105–108.

Leclerc, J.-C., Lamotte, M., and Richard-Molard, J., 1949, Niveaux et cycles d'érosion dans le massif du Nimba (Haute Guinée Française): Comptes Rendus de l'Académie des Sciences, Paris, v. 228, p. 1510–1512.

Leclerc, J.-C., Richard-Molard, J., Lamotte, M., Rougerie, G., and Portères, R., 1955, La réserve naturelle du Mont Nimba, fasc. 3, La chaîne du Nimba: Essai géographique: Mémoire de l'Institut Français d'Afrique Noire, v. 43, p. 1–271.

Le Maréchal, A., 1966, Contribution à l'étude des plateaux batékés. Géologie, géomorphologie, hydrogéologie. Brazzaville, Rapport ORSTOM, v. 137, 43 p.

Lepersonne, Y., 1960, Quelques problèmes de l'histoire géologique de l'Afrique du sud du Sahara, depuis la fin du Carbonifère: Bulletin de la Société Belge de Géologie, v. 84, p. 21–85.

Le Pichon, X., 1984, La tectonique des plaques: une tectonique globale de la lithosphère: Bulletin de la Société Géologique de France, v. 26, p. 345–359.

Leroux, M., 1996, La dynamique du temps et du climat: Paris, Masson, 310 p.

Leturmy, P., Lucazeau, F., and Brigaud, F., 2003, Dynamic interactions between the Gulf of Guinea passive margin and the Congo River drainage basin. 1. Morphology and mass balance: Journal of Geophysical Research, v. 108, doi: 10.1029/2002JB001927

Li, A., and Burke, K., 2006, Upper mantle structure of southern Africa from Rayleigh wave tomography: Journal of Geophysical Research, v. 111, p. B10303, doi: 10.1029/2006JB004321

Lithgow-Bertelloni, C.A., and Richards, M.A., 1995, The dynamics of Cenozoic plate driving forces: Geophysical Research Letters, v. 22, p. 1317–1320, doi: 10.1029/95GL01325.

Lithgow-Bertelloni, C.A., and Silver, P.G., 1998, Dynamic topography, plate-driving forces and the African superswell: Nature, v. 395, p. 269–272, doi: 10.1038/26212.

Lliboutry, L., 1982, Tectonophysique et géodynamique: Paris, Masson, 340 p.

Lucazeau, F., Brigaud, F., and Leturmy, P., 2003, Dynamic interactions between the Gulf of Guinea passive margin and the Congo River drainage basin. 2. Isostasy and uplift: Journal of Geophysical Research, v. 108, doi: 10.1029/2002JB001928

Lunde, G., Aubert, K., Lauritzen, O., and Lorange, E., 1992, Tertiary uplift of the Kwanza basin in Angola, in Curnelle, R., ed., Géologie africaine: Boussens, France, Elf Aquitaine, Colloque Géologique de Libreville, Recueil des Communications, p. 99–117.

Macgregor, D.S., Robinson, J., and Spear, G., 2003, Play fairways of the Gulf of Guinea transform margin, in Arthur, T.J., Macgregor, D.S., and Cameron, N.R., eds., Petroleum geology of Africa: New themes and developing technologies: Geological Society [London] Special Publication 207, p. 131–150.

Marker, M.E., and Brook, G.A., 1970, Echo Cave: A tentative Quaternary chronology of the eastern Transvaal: Johannesburg, Department of Geography, University of the Witwatersrand, Environmental Studies, v. 3, 38 p.

Marton, L.G., Tari, G.C., and Lehmann, C.T., 2000, Evolution of the Angolan Passive Margin, West Africa, with emphasis on post-salt structural styles, in Mohriak, W., and Talwani, M., eds., Atlantic rifts and continental margins: Washington, D.C., American Geophysical Union, p. 129–149.

Marzoli, A., Renne, P.R., Piccirillo, E.M., Ernesto, M., Bellieni, G., and De Min, A., 1999, Extensive 200 million-year-old continental flood basalts of the central Atlantic magmatic province: Science, v. 284, p. 616–618, doi: 10.1126/science.284.5414.616.

Massengo, A., 1970, Contribution à l'étude stratigraphique, sédimentologique et minéralogique de la série Plio-pléistocène du bassin côtier du Congo Brazzaville [Ph.D. thesis]: Talence, France, University of Bordeaux 1, 150 p.

Mateer, N.J., et al., 1992, Correlation of nonmarine Cretaceous strata of Africa and the Middle East: Cretaceous Research, v. 13, p. 273–318, doi: 10.1016/0195-6671(92)90003-9.

Matmon, A., Bierman, P., and Enzel, Y., 2002, Pattern and tempo of great escarpment erosion: Geology, v. 30, p. 1135–1138, doi: 10.1130/0091-7613(2002)030<1135:PATOGE>2.0.CO;2.

Matthews, A., Lawrence, S.R., Mamad, A.V., and Fortes, G., 2001, Mozambique basin may have bright future under new geological interpretations: Oil and Gas Journal, 2 July, p. 70–76.

Mayer, J.J., 1985, Interpretation of heavy mineral patterns in sandy soils of the region drained by the upper reaches of the Harts River: Transactions of the Geological Society of South Africa, v. 88, p. 27–32.

McConnell, R.B., 1955, The erosion surfaces of Uganda: Colonial Geology and Mineral Resources, v. 5, p. 425–428.

McFarlane, M.J., 1976, Laterite and landscape: London, Academic Press, 151 p.

McFarlane, M.J., 1991, Laterite and landscape: Journal of African Earth Sciences, v. 12, p. 267–282, doi: 10.1016/0899-5362(91)90076-B.

McKenzie, D., and Weiss, N., 1975, Speculation on the thermal and tectonic history of the Earth: Geophysical Journal of the Royal Astronomical Society, v. 42, p. 131–174.

McMillan, I.K., 1990, Foraminiferal biostratigraphy of the Barremian to Miocene rocks of the Kudu 9A-1, 9A-2 and 9A-3 boreholes: Communications of the Geological Survey of Namibia, v. 6, p. 23–29.

McMillan, I.K., 2003, Foraminiferally defined biostratigraphic episodes and sedimentation pattern of the Cretaceous drift succession (Early Barremian to Late Maastrichtian) in seven basins on the South African and southern Namibian continental margin: South African Journal of Science, v. 99, p. 537–576.

Menzies, M.A., Klemperer, S.L., Ebinger, C.J., and Baker, J., 2002, Characteristics of volcanic rifted margins, in Menzies, M.A., Klemperer, S.L., Ebinger, C.J., and Baker, J., eds., Volcanic rifted margins: Geological Society of America Special Paper 362, p. 1–14.

Merla, G., and Minucci, E., 1938, Missione geologica nel Tigre, Volume 1, La serie dei terreni: Rome, Reale Academica d'Italia, Centro Studi per l'Africa Orientale Italiana, v. 3, 363 p.

Michel, P., 1973, Les bassins des fleuves Sénégal et Gambie. Étude géomorphologique: Paris, Éditions ORSTOM, Mémoire 63, 752 p.

Mitchell, R.H., 1986, Kimberlites: Mineralogy, geochemistry, petrology: New York, Plenum Press, 442 p.

Moody, R.T.J., and Sutcliffe, P.J.C., 1991, The Cretaceous deposits of the Iullemmeden Basin of Niger, central West Africa: Cretaceous Research, v. 12, p. 137–157, doi: 10.1016/S0195-6671(05)80021-7.

Morin, S., 1989, Hautes terres et bassins de l'ouest du Cameroun: Étude géomorphologique [Ph.D. thesis]: Talence, University of Bordeaux 3, 1190 p.

Morley, C.K., Westcott, W.A., Stone, D.M., Harper, R.M., Wigger, S.T., and Karanja, F.M., 1992, Tectonic evolution of the Northern Kenyan rift:

Journal of the Geological Society [London], v. 149, p. 333–348, doi: 10.1144/gsjgs.149.3.0333.

Mortelmans, G., 1964, À propos de la présence au Katanga de cailloux éolisés dans le conglomérat de base des "grès polymorphes": Bulletin de la Société Belge de Géologie, v. 55, p. 220–228.

Mougamba, R., 1999, Chronologie et architecture des systèmes turbiditiques cénozoïques du prisme sédimentaire de l'Ogooué (Marge Nord-Gabon) [Ph.D. thesis]: Villeneuve d'Ascq, France, University of Lille 1, 185 + 100 p.

Mutwewingabo, B., 1989, Genèse, caractéristiques et contraintes des sols acides à horizon sombre de profondeur de la région de haute altitude du Rwanda, in Soltrop 89: Actes du premier séminaire franco-africain de pédologie tropicale, Lomé, 6–12 February 1989: Paris, Éditions ORSTOM, p. 353–385.

Needham, R.S., 1982, East Alligator, Northern Territory: Geological map and commentary: Canberra, Bureau of Mineral Resources, Geology and Geophysics, scale, 1:100 000, 27 p.

Nichol, J.E., 1998, Quaternary climate and landscape development in West Africa: Evidence from satellite images: Zeitschrift für Geomorphologie, v. 42, p. 329–347.

Nyblade, A.A., and Robinson, S.W., 1994, The African superswell: Geophysical Research Letters, v. 21, p. 765–768, doi: 10.1029/94GL00631.

Ollier, C.D., 1960, The inselbergs of Uganda: Zeitschrift für Geomorphologie, v. 4, p. 43–52.

Ollier, C.D., 1981, Tectonics and landforms: Harlow, Longman, 324 p.

Ollier, C.D., and Marker, M.E., 1985, The Great Escarpment of southern Africa: Zeitschrift für Geomorphologie, Suppl.-Bd., v. 54, 37–56.

Ollier, C.D., and Pain, C.F., 1997, Equating the basal unconformity with the palaeoplain: A model for passive margins: Geomorphology, v. 19, p. 1–15, doi: 10.1016/S0169-555X(96)00048-7.

Ollier, C.D., Lawrance, C.J., Beckett, P.H.T., and Webster, R., 1969, Land systems of Uganda: Terrain classification and data storage: Oxford-MEXE Report, v. 959, 234 p.

Omar, G.I., Steckler, M.S., Buck, W.R., and Kohn, B.P., 1989, Fission-track analysis of basement apatites at the western margin of the Gulf of Suez rift, Egypt: Evidence for synchroneity of uplift and subsidence: Earth and Planetary Science Letters, v. 94, p. 316–328, doi: 10.1016/0012-821X(89)90149-0.

Orme, A.R., 1973, Barrier and lagoon systems along the Zululand coast, South Africa, in Coates, D.R., ed., Coastal geomorphology: Binghamton, State University of New York Press, p. 181–217.

Orme, A.R., 2005, Africa: Coastal geomorphology, in Schwartz, M.L., ed., Encyclopaedia of coastal science: Dordrecht, Springer, p. 9–21.

Pallister, J.W., 1960, Erosion cycles and associated surfaces of Mengo District: Buganda: Overseas Geology and Mineral Resources, v. 8, p. 26–36.

Parrish, J.T., Ziegler, A.M., and Scotese, C.R., 1982, Rainfall patterns and the distribution of coals and evaporites in the Mesozoic and Cenozoic: Palaeogeography, Palaeoclimatology, Palaeoecology, v. 40, p. 67–101, doi: 10.1016/0031-0182(82)90085-2.

Partridge, T.C., 1997, Late Neogene uplift in eastern and southern Africa and its paleoclimatic implications, in Ruddiman, W.F., ed., Tectonic uplift and climatic change: New York, Plenum Press, p. 63–86.

Partridge, T.C., 1998, Of diamonds, dinosaurs and diastrophism: 150 million years of landscape evolution in southern Africa: South African Journal of Geology, v. 101, p. 167–184.

Partridge, T.C., and Maud, R.R., 1987, Geomorphic evolution of southern Africa since the Mesozoic: South African Journal of Geology, v. 90, p. 179–208.

Partridge, T.C., and Maud, R.R., 2000, Macro-scale geomorphic evolution of southern Africa, in Partridge, T.C., and Maud, R.R., eds., The Cenozoic of southern Africa: Oxford, UK, Oxford University Press, p. 3–19.

Pascual, J.-F., 1988, Les sols actuels et les formations superficielles des crêtes nord-est du Nimba (Guinée): Contribution à l'étude géomorphologique du Quaternaire de la chaîne: Paris, Cahiers ORSTOM, ser. Pédologie, v. 24, p. 137–162.

Patterson, S.H., Kurtz, H.F., Olson, J.C., and Neely, C.L., 1986, World bauxite resources: U.S. Geological Survey Professional Paper 1076-B, 151 p.

Penck, A., and Brückner, E., 1901–1909, Die Alpen im Eiszeitalter (3 vols.): Leipzig, Tauchnitz, 716 p.

Petit, C., Fournier, M., and Gunnell, Y., 2007, Tectonic and climatic controls on rift escarpments: Erosion and flexural rebound of the Dhofar passive margin (Gulf of Aden, Oman): Journal of Geophysical Research, v. 112, doi: 10.1029/2006JB004554

Petit, M., 1982, Essai de cartographie mondiale du cuirassement: Travaux et Documents du Centre de Géographie Tropicale, Bordeaux, v. 1, p. 1–15.

Petit, M., 1985, A provisional world map of duricrust, in Douglas, I., and Spencer, T., eds., Environmental change and tropical geomorphology: London, George Allen and Unwin, p. 269–279.

Petit, M., 1994, Les grands traits morphologiques de l'Afrique centrale atlantique, in Schwartz, D., and Lafranchi, R., eds., Paysages quaternaires de l'Afrique centrale atlantique: Paris, Éditions ORSTOM, p. 20–30.

Petroleum Agency of South Africa, 2003, South Africa: Petroleum exploration opportunities: http://www.petroleumagencysa.com (24 January 2006).

Pias, J., 1970, Les formations sédimentaires tertiaires et quaternaires de la cuvette tchadienne et les sols qui en dérivent: Paris, Éditions ORSTOM, Mémoire 43, 407 p.

Pik, R., Marty, B., Carignan, J., and Lavé, J., 2003, Stability of the Upper Nile drainage network (Ethiopia) deduced from (U-Th)/He thermochronometry: Implications for uplift and erosion of the Afar plume dome: Earth and Planetary Science Letters, v. 215, p. 73–88, doi: 10.1016/S0012-821X(03)00457-6.

de Ploey, J., Lepersonne, J., and Stoops, G., 1968, Sédimentologie et origine des sables de la série des sables ocre et de la série des "Grès polymorphes" (système du Kalahari) au Congo occidental: Tervuren, Musée Royal de l'Afrique Centrale, Annales Géologiques, v. 61, 72 p.

Prell, W.L., and Niitsuma, L., 1989, Introduction, background, and major objectives for ODP Leg 117 (Western Arabian Sea) in search of ancient monsoons, in Proceedings of the Ocean Drilling Program, Initial Reports, Volume 117: College Station, Texas, Ocean Drilling Program, p. 5–9.

Raab, M., Brown, R., Gallagher, K., Weber, K., and Gleadow, A., 2005, Denudational and thermal history of the Early Cretaceous Brandberg and Okenyenya igneous complexes on Namibia's Atlantic passive margin: Tectonics, v. 24, p. TC3006, doi: 10.1029/2004TC001688.

Redfield, T.F., Osmundsen, P.T., and Hendriks, B.W.H., 2005, The role of fault reactivation and growth in the uplift of western Fennoscandia: Journal of the Geological Society [London], v. 162, p. 1013–1030, doi: 10.1144/0016-764904-149.

Reyment, R.A., and Dingle, R.V., 1987, Palaeogeography of Africa during the Cretaceous period: Palaeogeography, Palaeoclimatology, Palaeoecology, v. 59, p. 93–116, doi: 10.1016/0031-0182(87)90076-9.

Roger, J., Platel, J.-P., Cavelier, C., and Bourdillon-de-Grisac, C., 1989, Données nouvelles sur la stratigraphie et l'histoire géologique du Dhofar (Sultanat d'Oman): Bulletin de la Société Géologique de France, v. 2, p. 265–277.

Rossi, G., 1980, Tectonique, surfaces d'aplanissement et problèmes de drainage au Rwanda-Burundi: Revue de Géomorphologie Dynamique, v. 3, p. 81–100.

Ruhe, R.V., 1956, Landscape evolution in the High Ituri, Belgian Congo: Brussels, Publications de l'Institut National pour l'Étude Agronomique du Congo, ser. Scientifique, v. 66, 92 p.

Rust, J.D., and Summerfield, M.A., 1990, Isopach and borehole data as indicators of rifted margin evolution in southwestern Africa: Marine and Petroleum Geology, v. 7, p. 277–287, doi: 10.1016/0264-8172(90)90005-2.

Saggerson, E.P., and Baker, B.H., 1965, Post-Jurassic erosion surfaces in eastern Kenya and their deformation in relation to rift structure: Quarterly Journal of the Geological Society [London], v. 121, p. 51–72.

Sahagian, D.L., 1988, Epeirogenic motion of Africa as inferred from Cretaceous shoreline deposits: Tectonics, v. 7, p. 125–138.

Saïd, R., Sabet, A.H., Zalata, A.A., Teniakov, V.A., and Pokryshkin, V.I., 1976, A review of theories of the geological distribution of bauxite and their application for bauxite prospecting in Egypt: Annals of the Geological Survey of Egypt, v. 6, p. 6–31.

Salman, G., and Abdula, I., 1995, Development of the Mozambique and Ruvuma sedimentary basins, offshore Mozambique: Sedimentary Geology, v. 96, p. 7–41, doi: 10.1016/0037-0738(95)00125-R.

Schmitt, J.-M., 1999, Weathering, rainwater and atmosphere chemistry: Example and modelling of granite weathering in present conditions, in a CO_2-rich, and in an anoxic palaeoatmosphere, in Thiry, M., and Simon-Coinçon, R., eds., Palaeoweathering, palaeosurfaces and related continental deposits: International Association of Sedimentologists Special Publication 27, p. 21–41.

Scholtz, A., 1983, The palynology of the upper lacustrine sediments of the Arnot Pipe, Banke, Namaqualand: Annals of the South African Museum, v. 95, p. 1–109.

Schwarz, T., 1994, Ferricrete formation and relief inversion: An example from central Sudan: CATENA, v. 21, p. 257–268, doi: 10.1016/0341-8162(94)90016-7.

Scotese, C.R., and Sager, W.W., 1988, Mesozoic and Cenozoic plate tectonic reconstructions: Tectonophysics, v. 155, p. 27–48, doi: 10.1016/0040-1951(88)90259-4.

Ségalen, P., 1967, Les sols et la géomorphologie du Cameroun: Paris, Cahiers de l'ORSTOM, ser. Pédologie, v. 2, p. 137–187.

Ségalen, P., 1995, Les sols ferrallitiques et leur répartition géographique. 3. Les sols ferrallitiques en Afrique et en Extrême-Orient, Australie et Océanie, Études et Thèses: Paris, Éditions ORSTOM, 201 p.

Şengör, A.M.C., 2001, Elevation as an indicator of mantle-plume activity, in Ernst, R.E., and Buchan, K.L., eds., Mantle plumes: Their identification through time: Geological Society of America Special Paper 352, p. 183–225.

Sepulchre, P., Ramstein, G., Fluteau, F., Schruster, M., Tiercelin, J.-J., and Brunet, M., 2006, Tectonic uplift and eastern Africa aridification: Science, v. 313, p. 1419–1423, doi: 10.1126/science.1129158.

Séranne, M., 1999, Early Oligocene turnover on SW African continental margin: A signature of Tertiary Greenhouse to Icehouse transition?: Terra Nova, v. 11, p. 135–140, doi: 10.1046/j.1365-3121.1999.00246.x.

Shipboard Scientific Party, 1999, Leg 178 summary: Antarctic glacial history and sea level change, in Barker, P.F., Camerlenghi, A., Acton, G.D., et al., eds., Proceedings of the Ocean Drilling Program, Initial Reports, Volume 178: College Station, Texas, Ocean Drilling Program, p. 1–60.

Shipboard Scientific Party, 2000, Leg 188 Preliminary Report: Prydz Bay-Cooperation Sea, Antarctica: Glacial history and paleoceanography: Ocean Drilling Program Preliminary Report, Volume 188 [Online], 79 p. Available from http://www-odp.tamu.edu/publications/prelim/188_prel/188prel.pdf (26 January 2008).

Smith, R.M.H., 1986, Sedimentation and palaeoenvironments of Late Cretaceous crater-lake deposits in Bushmanland, South Africa: Sedimentology, v. 33, p. 369–386, doi: 10.1111/j.1365-3091.1986.tb00542.x.

Spengler, A., de Castelain, J., Cauvin, J., and Leroy, M., 1966, Le bassin secondaire-tertiaire du Sénégal, in Bassins sedimentaires du littoral Africain: Symposium of the Association des Services Géologiques Africains, New Delhi, 1964, vol. 1: Paris, Association des Services Géologiques Africains, p. 80–97.

Spiegel, C., Kohn, B.P., Belton, D.X., and Gleadow, A.J.W., 2007, Morphotectonic evolution of the central Kenya rift flanks: Implications for Late Cenozoic environmental change in East Africa: Geology, v. 35, p. 427–430, doi: 10.1130/G23108A.1.

Stas, M., 1959, Contribution à l'étude géologique et minéralogique des bauxites du NE du Mayombe: Bulletin de l'Académie Royale des Sciences Coloniales (Bruxelles), v. 5, p. 470–493.

Steckler, M.S., and ten Brink, U.S., 1986, Lithospheric strength variations as a control on new plate boundaries: Examples from the northern Red Sea region: Earth and Planetary Science Letters, v. 79, p. 120–132, doi: 10.1016/0012-821X(86)90045-2.

Stevenson, I.R., and McMillan, I.K., 2004, Incised valley fill stratigraphy of the Upper Cretaceous succession, proximal Orange Basin, Atlantic margin of southern Africa: Journal of the Geological Society [London], v. 161, p. 185–208.

Stratten, T., 1979, The origin of the diamondiferous alluvial gravels in the southwestern Transvaal: Geological Society of South Africa Special Publication 6, p. 219–228.

Summerfield, M., 1985a, Tectonic background to long-term landform development in tropical Africa, in Douglas, I., and Spencer, T., eds., Environmental change and tropical geomorphology: London, George Allen and Unwin, p. 281–294.

Summerfield, M.A., 1985b, Plate tectonics and landscape development on the African continent, in Morisawa, M., and Hack, J.T., eds., Tectonic geomorphology: Boston, Unwin Hyman, p. 27–51.

Summerfield, M.A., 1996, Tectonics, geology and long-term landscape development, in Adams, W.M., Goudie, A.S., and Orme, A.R., eds., The physical geography of Africa: Oxford, UK, Oxford University Press, p. 1–17.

Sys, C., 1972, Caractérisation morphologique et physico-chimique de profils types de l'Afrique centrale: Brussels, Publications de l'Institut National pour l'Étude Agronomique du Congo, hors-série, 497 p.

Tardy, Y., 1993, Pétrologie des latérites et des sols tropicaux: Paris, Masson, 461 p.

Tardy, Y., and Roquin, C., 1998, Dérive des continents, paléoclimats et altérations tropicales: Orléans, Éditions du Bureau des Recherches Géologiques et Minières, 473 p.

Tari, G., Molnar, J., and Ashton, P., 2003, Examples of salt tectonics from West Africa: A comparative approach, in Arthur, T.J., Macgregor, D.S., and Cameron, N.R., eds., Petroleum geology of Africa: New themes and developing technologies: Geological Society [London] Special Publication 207, p. 85–104.

Taylor, R.G., and Howard, K.W.F., 1998, Post-Palaeozoic evolution of weathered landsurfaces in Uganda by tectonically controlled deep weathering and stripping: Geomorphology, v. 25, p. 173–192, doi: 10.1016/S0169-555X(98)00040-3.

Teisserenc, P., and Villemin, J., 1989, Sedimentary basin of Gabon—Geology of oil systems, in Edwards, J.D., and Santogrossi, P.A., eds., Divergent/passive margin basins: American Association of Petroleum Geologists Memoir 48, p. 117–199.

ten Brink, U., and Stern, T., 1992, Rift flank uplifts and hinterland basins: Comparison of the Transantarctic mountains with the Great Escarpment of southern Africa: Journal of Geophysical Research, v. 97, p. 569–585.

Thomas, M.F., 1994, Geomorphology in the tropics: A study of weathering and denudation in low latitudes: Chichester, UK, John Wiley and Sons, 460 p.

Torsvik, T.H., Smethurst, M.A., Burke, K., and Steinberger, B., 2006, Large igneous provinces generated from the margins of the large low-velocity provinces in the deep mantle: Geophysical Journal International, v. 167, p. 1447–1460, doi: 10.1111/j.1365-246X.2006.03158.x

Twidale, C.R., 1997, The great age of some Australian landforms: Examples of, and possible explanations for, landscape longevity, in Widdowson, M., ed., Palaeosurfaces: Recognition, reconstruction and palaeoenvironmental interpretation: Geological Society [London] Special Publication 120, p. 13–23.

Tyson, P.D., and Preston-Whyte, R.A., 2000, The weather and climate of southern Africa (2nd edition): Oxford, UK, Oxford University Press, 396 p.

Uenzelmann-Neben, G., 1998, Neogene sedimentation history of Congo Fan: Marine and Petroleum Geology, v. 15, p. 635–650, doi: 10.1016/S0264-8172(98)00034-8.

Upchurch, G.R., Otto-Bliesner, B.L., and Scotese, C.R., 1999, Terrestrial vegetation and its effects on climate during the latest Cretaceous, in Barrera, E., and Johnson, C.C., eds., Evolution of the Cretaceous ocean-climate system: Geological Society of America Special Paper 332, p. 407–425.

Valeton, I., and Beissner, H., 1986, Geochemistry and mineralogy of the Lower Tertiary in situ laterites of Jos Plateau, Nigeria: Journal of African Earth Sciences, v. 5, p. 535–550, doi: 10.1016/0899-5362(86)90063-1.

Van der Beek, P., Mbede, E., Andriessen, P., and Delvaux, D., 1998, Denudation history of the Malawi and Rukwa rift flanks (East Africa Rift System) from apatite fission track thermochronology: Journal of African Earth Sciences, v. 26, p. 363–385, doi: 10.1016/S0899-5362(98)00021-9.

Van der Beek, P.A., Summerfield, M.A., Braun, J., Brown, R.W., and Fleming, A., 2002, Modelling post-break-up landscape development and denudational history across the southeast African (Drakensberg Escarpment) margin: Journal of Geophysical Research, v. 107, doi: 10.1029/2001JB000744.

van der Hilst, R.D., and Karason, H., 1999, Compositional heterogeneity in the bottom 1000 kilometers of Earth's mantle: Toward a hybrid convection model: Science, v. 283, p. 1885–1888, doi: 10.1126/science.283.5409.1885.

Vanderstappen, R., and Cornil, J., 1955, Note sur les bauxites du Congo septentrional: Bulletin de l'Académie Royale des Sciences Coloniales (Bruxelles), v. 1, p. 690–709.

van der Wateren, F.M., and Dunai, T.J., 2001, Late Neogene passive margin denudation history—Cosmogenic isotope measurements from the central Namib desert: Global and Planetary Change, v. 30, p. 271–307, doi: 10.1016/S0921-8181(01)00104-7.

Van Niekerk, H.S., Beukes, N.J., and Gutzmer, J., 1999a, Post-Gondwana pedogenic ferromanganese deposits, ancient soil profiles, African land surfaces and palaeoclimatic change on the Highveld of South Africa: Journal of African Earth Sciences, v. 29, p. 761–781, doi: 10.1016/S0899-5362(99)00128-1.

Van Niekerk, H.S., Gutzmer, J., Beukes, N.J., Phillips, D., and Kiviets, G.B., 1999b, An $^{40}Ar/^{39}Ar$ age of supergene K-Mn oxihydroxides in a post-Gondwana soil profile on the Highveld of South Africa: South African Journal of Science, v. 95, p. 450–454.

Van Wambecke, A., 1963, Carte des sols et de la végétation du Congo, du Rwanda et du Burundi: Brussels, Publications de l'Institut National pour l'Étude Agronomique du Congo, 67 p.

Veatch, A.C., 1935, The evolution of the Congo basin: Geological Society of America Memoir 3, 183 p.

Vogt, J., 1959, Aspects de l'évolution morphologique récente de l'Ouest africain: Annales de Géographie, no. 68, p. 193–206, doi: 10.3406/geo.1959.16306.

Walford, H.L., White, N.J., and Sydow, J.C., 2005, Solid sediment load history of the Zambezi Delta: Earth and Planetary Science Letters, v. 238, p. 49–63.

Walgenwitz, F., Pagel, M., Meyer, A., Maluski, H., and Monié, P., 1990, Thermo-chronological approach to reservoir diagenesis of the offshore Angola basin: A fluid inclusion, ^{40}Ar-^{39}Ar and K-Ar investigation: American Association of Petroleum Geologists Bulletin, v. 74, p. 547–563.

Walgenwitz, F., Richert, J.-P., and Charpentier, P., 1992, Southwest African plate margin; thermal history and geodynamical implications, in Poag, C.W., and de Graciansky, P.C., eds., Geologic Evolution of Atlantic Continental Rises: Van Nostrand Reinhold, New York, p. 20–45.

Ward, J.D., and Corbett, I., 1990, Towards an age for the Namib: Transvaal Museum Monographs, v. 7, p. 17–26.

Wayland, E.J., 1933, The peneplain of East Africa: The Geographical Journal, v. 82, p. 95, doi: 10.2307/1786554.

Wayland, E.J., 1934, The peneplain of East Africa: The Geographical Journal, v. 83, p. 79, doi: 10.2307/1786689.

Whiting, B.M., Karner, G.D., and Driscoll, N.W., 1994, Flexural and stratigraphic development of the West Indian continental margin: Journal of Geophysical Research, v. 99, p. 13,791–13,811, doi: 10.1029/94JB00502.

Wilkinson, M.J., 2004, Megafans and continental sedimentation: New concepts in African exploration: "Africa Upstream 2004" conference, Cape Town, 6–8 October 2004, PowerPoint presentation,

Willis, B., 1936, East African plateaus and rift valleys: A study in comparative seismology: Washington, D.C., Carnegie Institution Publication 470.

Winn, R.D., Steinmetz, J.C., and Kerekgyarto, W.L., 1993, Stratigraphy and rifting history of the Mesozoic-Cenozoic Anza Rift Kenya: American Association of Petroleum Geologists Bulletin, v. 77, p. 1989–2005.

Wysick, P., Klitsch, E., Jas, C., and Reynolds, O., 1990, Intracratonal sequence development and structural control of Phanerozoic strata in Sudan: Berliner Geowissenschftliche Abhandlungen, ser. A., v. 120, p. 45–86.

Young, R.W., 1992, Structural heritage and planation in the evolution of landforms in the East Kimberley: Australian Journal of Earth Sciences, v. 39, p. 141–151, doi: 10.1080/08120099208728011.

Zachos, J., Pagani, M., Sloan, L., Thomas, E., and Billups, K., 2001, Trends, rhythms, and aberrations in global climate 65 Ma to present: Science, v. 292, p. 686–693, doi: 10.1126/science.1059412.

Manuscript Accepted by the Society 15 November 2007